Praise for THE UNQUIET GHOST

"An exceptionally perceptive and honest book that sensitively attempts to do justice to those who lived under Stalinist tyranny and the following forty years of state-imposed silence." — *San Francisco Chronicle*

"The characters and the dramatic situations Mr. Hochschild encounters are nothing short of magnificent." — *New York Times Book Review*

"[Hochschild] combines the strengths of a practiced investigative reporter with those of a philosopher-historian with a sensitive moral compass and the spirit of an enlightened eighteenth-century gentleman . . . An illuminating excursion led by a highly qualified guide."
— *Los Angeles Times Book Review*

"A book that adds greatly to our grasp of the dreadful phenomena of Stalinism." — Robert Conquest, author of *Stalin: Breaker of Nations*

"The most accurate, in-depth, and interesting exploration of the Stalinist past by an American journalist."
— Janusz Bardach, author of *Man Is Wolf to Man: Surviving the Gulag*

"A brilliant and compelling idea . . . pursued with great determination and resourcefulness." — *History Today*

"A horrifying and wondrous book." — *In These Times*

"Hochschild's skills as an interviewer and the honesty of his own questioning make for a thoroughly compelling and original book."
— *New Statesman*

"Over the years a few courageous and dedicated writers have compiled an essential record of the Stalinist terror: Alexander Solzhenitsyn in novels, Nadezhda Mandelstam, Eugenia Ginzburg, and others in memoirs, and Robert Conquest in history books. Hochschild deserves a place among these illustrious names." — *Montreal Gazette*

"Hochschild's people, as well as his honesty and passion, make this unforgettable book essential for everyone concerned about history and human rights." — *Library Journal*

"A riveting and eloquent evocation of a country still haunted by the ghost of Stalin . . . As se━━━━━━━━━━━━━━━━━━━━khov: journalism raised to the lev━━━━━

The
Unquiet
Ghost

BOOKS BY ADAM HOCHSCHILD

Half the Way Home
A MEMOIR OF FATHER AND SON

The Mirror at Midnight
A SOUTH AFRICAN JOURNEY

The Unquiet Ghost
RUSSIANS REMEMBER STALIN

Finding the Trapdoor
ESSAYS, PORTRAITS, TRAVELS

King Leopold's Ghost
A STORY OF GREED, TERROR,
AND HEROISM IN COLONIAL AFRICA

The Unquiet Ghost

RUSSIANS REMEMBER STALIN

Adam Hochschild

A MARINER BOOK
HOUGHTON MIFFLIN COMPANY
BOSTON · NEW YORK

First Mariner Books edition 2003

Preface copyright © 2003 by Adam Hochschild
Copyright © 1994 by Adam Hochschild
All rights reserved

For information about permission to reproduce selections from this book,
write to Permissions, Houghton Mifflin Company, 215 Park Avenue South,
New York, New York 10003.

Visit our Web site: www.houghtonmifflinbooks.com.

Library of Congress Cataloging-in-Publication Data is available.
ISBN 0-618-25747-0 (pbk.)

Printed in the United States of America

QUM 10 9 8 7 6 5 4 3 2 1

Portions of this book first appeared, in somewhat different form, in the *Los Angeles Times,*
Mother Jones, the *New York Times Magazine, In These Times,* and *West* magazine.

Grateful acknowledgment is made for permission to reprint excerpts from the following
copyrighted works: "Destruction" by Nikolai Kluyev, trans. by Richard McKane, appear-
ing in *Index on Censorship.* Reprinted by permission of *Index on Censorship,* London.
"History of Russia" from *Resistance* by Victor Serge, trans. by James Brook. Copyright
© 1989 by James Brook. Reprinted by permission of City Lights Books. *Hope Against
Hope* by Nadezhda Mandelstam, trans. by Max Hayward. Copyright © 1970 by Atheneum Publishers. Reprinted with the permission of Atheneum Publishers, an imprint of
Macmillan Publishing Company. *Hope Abandoned* by Nadezhda Mandelstam, trans. by
Max Hayward. Copyright © 1973, 1974 by Atheneum Publishers. Reprinted with the
permission of Atheneum Publishers, an imprint of Macmillan Publishing Company.
Journey into the Whirlwind by Eugenia Semyonovna Ginzburg, trans. by Paul Stevenson
and Max Hayward. Copyright © 1967 by Arnoldo Mondadori Editore, Milan. English
translation copyright © 1967 by Harcourt Brace & Company. Reprinted by permission
of Harcourt Brace & Company. *Within the Whirlwind* by Eugenia Ginzburg. Copyright
© 1979 by Arnoldo Mondadori Editore, S.p.A., Milan. English translation copyright ©
1981 by Harcourt Brace & Company and William Collins Sons & Co., Ltd. Reprinted by
permission of Harcourt Brace & Company.
    Photograph credits: frontispiece, xvi, 93, 287, David King Collection; 1, UPI/Bett-
mann; 41, courtesy of Sergei Krasilnikov; 99, 134, 214, 225, Ellen Binder; 157, Adam
Hochschild; 195, Russian Pictorial Collection, Hoover Institution Archives; 233, Thomas
Sgovio Collection/Hoover Institution

Frontispiece photograph: Stalin in 1933, said to be signing a death sentence

For Janusz Bardach (1919–2002)

Contents

Preface

A few days after arriving in Moscow to begin the research for this book, I trudged through the snow to go shopping at a grocery store near our rented apartment. An older man with a lined, thoughtful face seemed to have the job of stocking the shelves—easy work, because there was not much food. He was sitting in a chair at the end of a deserted aisle reading Varlam Shalamov, the great short story writer of the *gulag*. "I've read that same book in English," I told him. We agreed that Shalamov was a great writer. "And it's important," the man said with much feeling, "that these things be *documented,* by someone who speaks with the authority of having been there." Probably, like so many millions of other Russians, he had seen someone from his own family vanish into the wintry darkness of the Siberian labor camps.

The Russia where I had come to live, in 1991, was then a country where only in the previous few years had it become possible to read those long-forbidden books at last, to look the past in the face, and to ask the question that obsessed Russians as much as it did me: How could the country that gave the world Tolstoy and Chekhov also give it the *gulag*? Everywhere I went, it seemed, people were thinking about this. In addition to the men and women I sought out to interview, I kept stumbling into other unplanned conversations about Russia's Stalinist past, from the one in that grocery store to one recounted in the following pages, which had to be carried on in shouts, above the roar of jet engines, in a helicopter's cockpit.

It is now more than a decade since I spent half a year in Russia, first in Moscow and then traveling all the way to the country's Pacific coast, finding the people whose voices are heard here. After I returned home, friends often asked me, Wasn't it depressing to live there for so many months, studying the blood-soaked Stalin years? But it was not. On the contrary, horrible as that era was, this made it all the more inspiring to spend time with people who were bravely seeking to grapple with that dark time, to understand it deeply, to remember all that the authorities had wanted forgotten— and to try to make sure, by means of that remembering, that history does not repeat itself.

I hope you will find it as moving as I did to meet the people in this book: veterans of the *gulag,* men and women whose parents are buried in mass graves, a daughter struggling to reconcile her love for her father with the hundreds of death warrants he signed, two most unusual former secret policemen, and a woman who, as a teenager half a century ago, had courage matched by no adult around her.

Like the characters in the great nineteenth-century Russian novels, I found many of these people larger than life—or at least grappling with moral issues that seemed larger than those that confront most of us in the West. Few of us ever face the choice of whether to testify against a friend or neighbor or be shot ourselves. And few of us today have to judge a parent or grandparent who made such a decision. Yet these issues are all too familiar in too many parts of the world. Since this book was written, Rwanda and the former Yugoslavia have erupted in horrendous mass murder, where often it was friends and neighbors who did the killing or who passively stood by. The questions raised by human behavior in Stalin's Soviet Union still face us in the twenty-first century.

All Russians struggling to come to grips with the Stalin era know what Thomas Hardy meant when he wrote that "if way to the Better there be, it exacts a full look at the Worst." It was wanting to see just how people in Russia were taking that full look that took me there. I was extremely lucky, I realize now, to have gone when I did. A half decade earlier, the lid was screwed on tight and even the most oblique "look at the Worst" was impossible. No ideas, no history, no writers outside the gray Communist Party orthodoxy could be read or talked about in public. By a year or two after my stay in the Soviet Union during its final year as one country, people still enjoyed the freedom to explore the past, but were overwhelmed by the collapse of the economy, official corruption, and the shocking rise in organized crime. It is hard to focus on injustice in the past when you've lost your job and someone has just been murdered on the street outside.

And so 1991 turned out to be the right moment for my journey of exploration. It was a time when mass graves had been newly opened, and I was able to walk through several of them, seeing, in one, skull after skull with a bullet hole through it. It was a time when it was finally possible to see the old *gulag* camps, and I will never forget standing in the ruins of one of them, Butugychag, a place so cold and remote and surrounded by barren snow-streaked rock hills that it seemed like another planet. Even there, in that desolate moonscape with nothing but snowfields to escape to for dozens of miles, even there, the camp had an internal prison with thick stone walls and cross-hatched iron bars on the windows. Above all, it was a time when people who had survived such camps and the era that produced them were eager to tell their stories. It would be harder to gather such stories today, because so many of those who spoke to me were in their seventies or eighties and are now dead.

In another way, also, I was unexpectedly lucky in my timing. The very month I arrived in Russia, the government lifted prohibitions that for decades had placed huge swaths of the country off-limits to foreigners. This meant that in several places I visited in Siberia, I was the first American or Western European whom anyone there had ever seen. I could feel people's eyes on me as I got off the plane, almost as if I were a member of another race entirely; I fancied I could hear them thinking, "Well, he doesn't look that different: he's got two arms like us, and two legs like us . . ." The novelty of meeting their first Westerner made many people particularly eager to tell their stories. I was the first witness from another world.

After this book was published in the United States and Britain, it came out in translation in several countries in Europe and a portion of it appeared in a Russian magazine published by former *gulag* prisoners. It has been moving to read the letters, poetry, and e-mail people have sent me. Some have been survivors of the Holocaust, the war in Bosnia, and other catastrophes, who have found in the moral questions facing people in the Stalin-era Soviet Union some echo of their own experience. And some have been survivors of Stalin's own particular holocaust—a reminder of how widely his network of terror reached, and how much it has seared the lives of later generations throughout the world.

From Israel, an artist wrote, "My father, Benjamin Levinstein, is buried ... in the Lower At-Uryakh camp, the one you flew over. ... I enclose a photo of two pictures I have painted and dedicated to the memory of my father's martyrdom." From Germany, a young woman wrote about her childhood in one of the cities I visited, Karaganda; a friend translated the letter for me: "After reading your chapter 4, I remembered funeral processions in Karaganda when I observed totally unfamiliar onlookers weeping, convulsively weeping. At age nine or ten I mustered courage to ask my mother and aunts about this. The ... answer was: the weeping is not for the individual dead person; we weep because everybody (in Karaganda really everybody) has suffered so much loss and pain that the opportunity to weep is a release, an outlet."

From Kansas, a woman wrote, "For the past fifty-five years, I have been struggling with wanting to know who I am. At the age of three or four years, I remember knowing my Father was shot. About that time, my Mother was taken from me and sent to Siberia. I survived and at the age of fifteen came to America. ... At the age of fifty-nine, I finally found the only family member that survived living in the Ukraine—my Brother, Igor. He does not speak English and I can no longer speak Russian."

There were also stories that held triumph as well as pain. One

day soon after the book appeared my telephone rang and a booming, accented voice said, "I rrrrrread your boooooook!" He paused.
"And I was *there*." The voice belonged to Dr. Janusz Bardach, who,
it turned out, had spent four years in the camps of Kolyma, the
deadliest corner of the *gulag*, to which the last fifty of these pages are
devoted. Bardach had not only survived, he had survived even an
internal camp punishment cell like the ones I had seen, its floor deliberately covered in freezing water. Then he had eventually made
his way to the United States and become a renowned reconstructive
surgeon, the inventor of some of the standard procedures used to repair cleft lips and palates throughout the world. Stalin's regime took
his freedom and very nearly his life; he spent his career giving children back their faces. We became dear friends, and before his recent death, he finished two extraordinary volumes of memoirs of
Stalin's Russia, the first about his time in Kolyma, *Man Is Wolf to
Man: Surviving the Gulag*.

As both perpetrators and surviving victims die, how much does
Stalin's legacy still shadow Russia today? This is, after all, a country
that in 1999 installed as its leader Vladimir Putin, a man who spent
much of his career in the secret police. He has put many of his former comrades in high government positions. Furthermore, Russia
has fought several rounds of a brutal war against nationalists in
Chechnya, and there are separatist stirrings in other regions of the
country. Wars are never healthy for democracy and civil liberties,
especially when both are, as is the case in Russia, incomplete and
fragile to begin with. Another economic collapse could also easily
trigger Russia's traditional search for scapegoats, foreign and do-
mestic.

Weighing against these forces, of course, is an opposite one: the
deep, urgent desire by tens of millions of Russians, high and low, in
government and out, to become politically and economically more
integrated into Europe. In the long run, I suspect, the second set of
pressures will be the more powerful. But despite all the trappings of
courts and elections, no one should have any illusions that the country's path will be a smooth and easy one. It was, of all people, a secret police major who told me (page 160) that he thought it would

take two generations for the Russian bureaucracy to completely outgrow its authoritarian habits.

In the long run, the worst legacy of Stalinism is not the tendency to throw someone in prison first and ask questions later. It is the habits of mind, among those not directly affected, that make such acts even thinkable. For the quarter-century of Stalin's rule, and to a lesser extent before and after, voicing any opposition to the government could get you dispatched to a minus-40-degree camp in Siberia for years of hard labor. A vastly greater number of people were shot or imprisoned *without* voicing any opposition at all. After such an experience, the message survivors have always given their children is *Keep your head down.* Over time, stark fear turned into passivity. But this also is not fertile soil for the culture of democracy and all that goes with it: a bold, diverse press; education not by rote; untrammeled labor unions; community organizations of all kinds; citizens' groups that feel free to agitate for any cause, wise or foolish. Despite the heritage of fear and passivity, many brave and thoughtful Russians have been looking deeply into their country's past, and that is the story I have tried to tell here. It is people like them who will be the backbone of whatever civil society evolves in Russia, and I feel honored to have spent time in their company.

October 2002

Major Events in 20th-Century Russian and Soviet History

1894–1917: Reign of Tsar Nicholas II.

1903: At a congress of revolutionaries in exile in Western Europe, Russian Social Democratic Party splits; Lenin leads the more militant Bolshevik faction.

1904–1905: Russo-Japanese War, ending in humiliating defeat for Russia.

1905: Revolutionary upheavals; troops fire on workers marching in St. Petersburg. Nation-wide general strike. Tsar reluctantly agrees to constitution and legislature with limited powers.

1906–1907: Tsar keeps most authority, reduces legislative representation of non-Russians, workers, and peasants.

1914: Russia goes to war against Germany, Austria-Hungary, and Turkey.
St. Petersburg, the capital, Russifies its name to Petrograd.

1914–17: Ill-equipped and incompetently led, 7.9 million Russian troops are killed, wounded, or taken prisoner in World War I. Huge losses of territory to the Germans.

1917: March (New Style calendar): Strikes, food riots sweep Petrograd; troops mutiny. Tsar abdicates and is arrested. Provisional Government forms.
April–May: Lenin, Trotsky, and other revolutionary leaders return from exile abroad or in Siberia.

November: Calling for "peace, land, and bread," Bolsheviks, supported by soldiers, sailors, and Petrograd workers, storm the Winter Palace, arrest Provisional Government leaders, and seize power. Lenin, Trotsky, Stalin, and Kamenev are among the leaders of the new regime.

November: Russia's first national democratic election. Bolsheviks win only a quarter of Constituent Assembly seats.

December: Bolsheviks form secret police under Felix Dzerzhinsky.

1918: January: Bolshevik troops break up Constituent Assembly. Government institutes forced labor for prisoners.

March: Peace treaty signed with Germany at Brest Litovsk. Germans occupy most of the Ukraine.

March: Bolsheviks rename themselves the Communist Party, move capital to Moscow.

July: Tsar Nicholas II and his family executed.

1918–20: Russian Civil War. Red Army led by Trotsky. Opposing Whites led mostly by former Tsarist generals. Both sides execute prisoners en masse, suffer high casualties. American, British, French, and Japanese troops seize major Russian ports, and arm and supply the Whites.

1920: Communists ban all other political parties.

1922–25: Rule of the "Triumvirate": Stalin, Kamenev, and Zinoviev.

1924: Lenin, ill for several years, dies. Petrograd is renamed Leningrad.

1926–29: Winning the struggle for succession, Stalin amasses all political power, and jails or exiles his opponents.

1929–33: All peasants forced onto collective farms. *Kulaks,* the better-off peasants, deported to Siberia and Central Asia. Food production drops catastrophically. Seven to ten million people starve.

1929: First Five-Year Plan inaugurates crash industrialization. Production quotas for all factories and collective farms.

1932: First large transports of prisoners sent to Kolyma region.

1933: More than 100,000 prisoners finish digging the White Sea Canal, the first huge project built with forced labor. Canal fills with silt, cannot hold large ships.

1936: First of three big Moscow show trials of former leaders. Kamenev, Zinoviev, and fourteen others found guilty of treason, shot. Many other trials, public and secret, follow.

1936–39: The Great Purge. Unparalleled wave of arrests. Majority of Party leadership and Red Army senior commanders shot. Key professions are decimated.

1935–41: More than 19 million Soviets arrested. Seven million of these shot; majority of the remainder die in the *gulag*.

1939: Soviet Union signs non-aggression pact with Nazi Germany.

1939–40: Under terms of the pact, Soviet troops occupy Latvia, Lithuania, Estonia, half of Poland, part of Romania.

1941: German troops attack the Soviet Union, overrun vast areas, reach outskirts of Moscow and Leningrad.

1941–45: Twenty-seven million Soviet soldiers and civilians die in World War II.

1945: All Soviet prisoners of war who survive Nazi captivity are promptly sent to Soviet prison camps.

1947–53: New waves of mass arrests. Many victims are Jews.

1953: Stalin dies.
Secret police chief Lavrenti Beria ousted and executed.

1954: Prisoners revolt in several major *gulag* areas.
Slowly, the government begins "rehabilitating" some Great Purge victims, posthumously declaring them innocent.

1956: Nikita Khrushchev, now the country's leader, denounces Stalin's crimes in a secret speech to a Party congress.
An estimated 7 to 8 million prisoners released from the *gulag*.

1962: Solzhenitsyn's *One Day in the Life of Ivan Denisovich* published in Russia; high point of cautious, short-lived "thaw" under Khrushchev.

1964: Khrushchev ousted.

1964–82: Leonid Brezhnev in power. De-Stalinization, rehabilitations halt; censorship tightens; harsher persecution of political and religious dissidents.

1985: Mikhail Gorbachev takes power. Introduces *glasnost,* economic reforms, moves to end Cold War.

1986–91: Censorship dissolves. Previously banned books and films appear. Rehabilitations resume. Deepening economic crisis. Central government authority weakens.

1991: Leningrad votes to return to the name St. Petersburg. Conservative military and Party leaders attempt coup, fail. Soviet Union dissolves; Russia and the fourteen other republics become separate countries.

1993: Some 140 people die in Moscow street fighting between conservative rebels and forces loyal to Boris Yeltsin. Pressure for regional autonomy intensifies throughout Russia.

Introduction:
The Great Silence

In the early 1980s, I was on a visit to Moscow with a delegation of American journalists, politicians, and academics. Under the glittering chandeliers of a pre-revolutionary merchant's mansion, we held a week of talks about Cold War issues with Soviet intellectuals and government officials. This was before *glasnost;* although some of the Russians were fairly prominent, none felt free to voice any real criticisms of official policy. As a result, the meetings were stupefyingly boring.

One day we were sitting at lunch with our Soviet hosts. The talk remained polite and stilted. Then an American at my table noticed an old friend walking through the restaurant. He waved at the man, who came over and chatted for a few minutes. The two laughed and reminisced about the last time they had seen each other, many years before, during the stormy street demonstrations at the Democratic National Convention in Chicago in 1968. "... Weren't you arrested that night too? What a scene in the jail! Remember? Tom Hayden was there, Dr. Spock was there, *everybody* was there. We sang all night! It was fantastic. . . ."

A young Russian was sitting at our table, one of the most suave and sophisticated members of the Soviet delegation, a man who subtly flaunted his privileged status by cracking an occasional anti-Soviet joke. But when he overheard these two Americans, inhabitants of a strange, blithe planet where people sang in jail, a wave of shock crossed his face. After their conversation was over, he said, softly but with great feeling, *"You must not joke about being in prison."*

It was a sudden reminder of the chasm between our own experience and that of people in Russia. Like tens of millions of others, I guessed, this man must have had someone from his family—a father, a grandfather, an aunt—sent to prison or shot during Stalin's time. He still could not talk about it openly, however, especially with a group of foreigners, where his job was to say how much the Soviet Union wanted peace. The talk around the table moved on to other subjects. But for a moment, in the unexpected intensity of his voice, I felt as if I had glimpsed the hidden part of an iceberg, before the water covered it again.

I have had other such glimpses on trips to Russia over the years, although before *glasnost* they were usually fleeting, and, if strangers were in earshot, it was hard to get someone to say much. For almost as long as I have been reading, I have been drawn to the submerged part of that iceberg: the history of the Soviet Union under Stalin. There are many reasons; one of them, surely, is the spectacle of domination and evil on such an epic scale. In few major nations in modern times has one man wielded such absolute power for so many years (Stalin ruled, after all, more than twice as long as Hitler). And perhaps no tyrant at any time so thoroughly tried to rearrange reality itself. Stalin's historians and encyclopedists rewrote the past to make him a great hero of the Russian Revolution and the greatest statesman of all time. His pet scientist, Trofim Lysenko, rewrote biology to eliminate genes and let acquired characteristics be inherited. Farm and factory managers rewrote production statistics to meet the high quotas the government demanded. Stalin dispatched some fifteen hundred writers to their deaths, and often rewrote the manuscripts of those who survived, marking his changes in red or green pencil ("I have added a few words on page 68," he wrote to the novelist Alexander Korneichuk. "It makes things clearer"). His ambition to rewrite history knew no limits. Nadezhda Krupskaya, Lenin's widow, eventually died of natural causes—a rare achievement in the Soviet Union of the 1930s—but Stalin threatened that if she did not stop criticizing him, the Communist Party might have to find Lenin a new widow.

In a recent piece by a Russian satirist, Stalin comes back to life, looks around, and asks who started all the changes.

"Khrushchev," is the answer.

"Execute him!" says Stalin.

"He's already dead."

"Execute him posthumously!" says Stalin.

Execution was the favored solution to every problem, including those caused by previous executions. When the national census showed that his reign of terror was shrinking the country's population, Stalin ordered the members of the census board shot. The new officials, not surprisingly, came up with higher figures. Between about 1929, when Stalin had vanquished his rivals and concentrated power in his hands, and his death in 1953, most historians now estimate that he had been directly responsible for the deaths of somewhere around 20 million people.

In a country where some archives are closed and others destroyed, and where uncounted bodies were sometimes bulldozed into mass graves, statistics are far from exact, and often in dispute. Most estimates are that between a third and half the dead perished in famines in the early 1930s. This was when the Soviet government forced peasants onto the new collective farms. In the process the authorities took over the land of the *kulaks,* the better-off peasants, which usually meant anyone who had two or more animals. Many *kulaks* were shipped off to Siberia in boxcars at the beginning of winter, and starved. These farmers had been the country's most productive, and starvation also claimed millions of other people who depended on the food that they had grown.

The remainder of those who died ended their lives in execution cellars and in the huge network of concentration camps. The *gulag* existed for most of Soviet rule, but mass arrests reached their height in the Great Purge, a nationwide frenzy of jailing and killing in the late 1930s. And here the numbers appear more definite.

Collective farm workers proclaiming the liquidation of the *kulaks*.

A confidential study of secret police documents ordered by Nikita Khrushchev reportedly found that between 1935 and 1941, the authorities arrested more than 19 million Soviets. Seven million of them were shot outright. The majority of the rest died of malnutrition or exposure, or sometimes in later waves of executions, in the far-flung camps of the *gulag*.

Unlike the Nazis who killed Jews, Gypsies, and other groups, the Turks who killed Armenians, or the Belgian colonizers who killed native inhabitants of the Congo, the executioners of the Great Purge did not choose their victims on an ethnic basis. Instead, the Purge was one of those moments in history—like the Spanish Inquisition or the reign of the guillotine in France—when a religious or political movement turns inward and begins devouring its own. Paradoxically, this crazed slaughter of imaginary enemies did not erupt in full force until the new Soviet Union's real enemies were no longer a threat. Surviving supporters of the old regime had long since left for Paris or New York. The *kulaks* were collectivized or dead. The arch dissenter Leon Trotsky was in exile in Mexico. Anyone openly opposed to Stalin at home was in prison.

Only then, strangely, did the Great Purge begin. It rolled through almost every Soviet ministry, university, factory, village, and apartment house. On the walls of overflowing prisons throughout the country was scrawled the word, *zachem?* (Why?) A bewildered world watched hundreds of top Red Army generals arrested and shot, and veteran revolutionaries accused at show trials in Moscow of being members of plots or conspiracies against the regime. To be a dutiful true believer was no protection. It was even a danger, for the vast apparatus of torture and death quickly swallowed up well over a million members of the Communist Party. Five successive First Secretaries of the Komsomol, the Party's youth organization, were shot. Thirteen successive secretaries of the Academy of Sciences in Kiev were arrested. The wave of mass arrests even swept up some twenty thousand members of the secret police itself. "Bravo!" the monarchist officer in Arthur Koestler's *Darkness at Noon* taps in code on his cell wall when he learns that the next cell contains a high-ranking Party official. "The wolves devour each other." The Great Purge is a metaphor for our time: to confront it fully is to grapple not only with the human capacity for cruelty to others, but also with our capacity for self-destruction.

For many Russians who lived through it, the era of Stalin remained the central trauma of their lives. The great poet Anna Akhmatova waited endlessly in lines outside Leningrad prisons for news of her husband and son:

> And if some day in this country
> They decide to raise a monument to me,

On only one condition will I consent
To this commemoration: if they do not place it
By the sea, where I was born . . .
Nor in the royal gardens . . .
But *here,* where for three hundred hours I stood,
And where they never unlocked the door for me.

That climate of cold, pervasive fear, of denunciations and confessions and midnight arrests, has long seemed to the survivors a
moral touchstone that revealed a person's fiber, and showed who
your true friends were. In a recent memoir about the arrest of her
father, the scientist Natalya Rappaport writes: "35 years have
passed since then, and in my life there have been many different
encounters, but still today when I meet new people I measure them
by the standards of those years, and I ask myself how they would
have acted."

Among what makes the Stalin period so riveting, then, is the
vastness of the suffering, the strange spectacle of self-inflicted genocide, and the way people's courage or cowardice was laid bare.
There is also one more thing that gives the era special meaning.
What happened in those years was not only evil but, in the original
sense of the word, tragic. For the famine, the *gulag,* and the millions of executions evolved not out of a movement that now seems
thoroughly sinister from the start, like the European conquest of
Africa or the ambitions of Hitler, but out of an event, the Russian
Revolution, which was celebrated by good people, enlightened people, progressive people all over the world. With its great gap between autocratic rich and illiterate poor, its prisoners, censorship,
and secret police, and its all-powerful Tsar, the old Russia cried out
for drastic change. To anyone who dreamed the great Utopian
dreams of justice and human betterment, there seemed no nation
that needed a revolution more.

The desire to eradicate tyranny and suffering is one side of the
Utopian impulse. All sorts of good ideas, from the abolition of slavery to equal rights for women, were first scorned as Utopian, then

gradually accepted. However, there is another, more hazardous facet of Utopianism: the faith that if only we make certain sweeping changes, then all problems will be solved. Most of us have felt, at one time or another, the appeal of a simple solution for life's difficulties. And one aspect of Marxism offered this: the belief that once people overthrew the social and economic system under which they lived, human character itself would be transformed. With the Russian Revolution, this transformation seemed at first to be taking place as predicted, and, miraculously, in Europe's most backward major nation. In contrast to various other imagined Utopias—the arrival of a Messiah, the world of the Noble Savage, proposals for communal farms—this one was actually happening, and on a huge scale. You could get on a ship or a train and go there.

The promised Utopia, of course, rapidly became quite the opposite. Even though it took a dozen years after the 1917 Revolution for Stalin's dictatorship to be complete, ominous signs increased throughout the 1920s: the swift suppression of all rival political parties, the rising power of the secret police, the ever-tightening censorship. Thus a visit to the new Soviet state was always a laboratory test of an idealistic traveler's ability to see clearly. Few did. The muckraking journalist Lincoln Steffens found, in his famous phrase, the future that worked. In the famine year of 1931, George Bernard Shaw discovered that Stalin was a fine fellow and that everyone in Russia had plenty to eat. An appalling number of other people who should have known better, from Theodore Dreiser to Julian Huxley to Beatrice and Sidney Webb, came home with tales of proud steelworkers and apple-cheeked milkmaids.

As we look back today at this generation of starry-eyed fellow travelers of some sixty years ago, it is easy to shake our heads at their naivete. But at the same time, I've long had the uncomfortable feeling that if I had come of age in those decades, I, too, would have welcomed the Russian Revolution with great hope. If you believed in social justice and equality, and in the rights of labor, women, and minorities, how splendid it would have been to brandish the Soviet

Union in the face of conservatives at home: the Russians are building a whole new world over there—and it works! Why can't we do the same?

In every way, the idea of communism embodied the hope of human transformation. It transcended the blind nationalism behind the appalling slaughter of World War I. It offered an alternative to the despair and unemployment of the Great Depression. It promised a sense of comradeship that would gradually spread to all countries of the earth. It gave intellectuals solidarity with the working class (and how especially easy it was to feel that bond with the working class of another, far-off country). Finally, it allowed you to enjoy that solidarity at the very same time you were in an enlightened vanguard that understood exactly how history was unfolding. How magic a combination!—to be a stalwart ally of the toiling masses and a member of an elite, all at once.

By the time of my own political generation, that of the 1960s, most of us had learned at least something from preceding decades, and were more likely to admire Orwell than Lenin. Still, some people briefly projected their dreams of the ideal society onto Cuba or China or North Vietnam. If you can imagine Utopia, it seems, there is a temptation to believe that somewhere, especially somewhere far away and difficult to get to, it already exists. Earlier in the century, when the regime that seized power in distant Russia was the very first of its kind, this temptation was overwhelming. And so part of what leads me back to that time, when so many people believed in the Russian Revolution's promise to end injustice forever, was the question of whether, had I been alive then, I would have been among them. And that leads in turn to the larger question posed by how both Russians and Westerners of the twenties and thirties idealized the Soviet Union: What makes for clear-seeing, and what makes for denial?

For Russians, the question of seeing clearly still applies, in different form, today. No nation easily faces a shameful period of its past. Before the 1970s, none of the exhibits at Colonial Williamsburg acknowledged that half the people in the original Williamsburg were slaves. Only in the 1980s did U.S. school textbooks admit that the arrival of Columbus and the westward expansion of the frontier was not good news for American Indians. Japanese schoolbooks still say little about Japan's brutal imperial conquests of the 1930s. A prominent group of revisionist historians in Germany claims that, in some ways, the Nazis weren't so bad after all.

In Germany, the job of facing the past is far from complete, but in Russia it has barely begun. Even long after the bloodshed was over, what made Russia remarkable was the scale of the silence. For decades after Stalin died in 1953, nearly all public discussion of his rule was strictly forbidden. Only for a brief interlude was this not so. Khrushchev's famous "secret speech" denouncing Stalin's crimes—rather tamely, given what was known—came in 1956, and only the Communist Party elite was privy to it. Over the next half dozen years a few books about the labor camp system appeared, like Alexander Solzhenitsyn's *One Day in the Life of Ivan Denisovich*. Stalin's body was quietly slipped out of Lenin's tomb, and the thousands of schools and factories and towns bearing Stalin's name were renamed after other people. But little public comment was allowed except in the vague, deadening Party rhetoric about "errors and excesses during the period of the personality cult." There was no real debate: too many people still in power had blood on their hands.

Soon the authorities screwed the lid on again, very tightly. Khrushchev was overthrown in 1964, and his secret speech itself was not published in the Soviet Union until more than thirty years after he gave it. *One Day in the Life of Ivan Denisovich* vanished from libraries. The government banned all Solzhenitsyn's other novels, as well as his epic dissection of the prison system, *The Gulag Archipelago*. Historians wrote about other periods. For three and a half decades after the dictator's death, not one new Soviet biography of Stalin, from any point of view, was allowed to be published. In the more than two dozen volumes of the edition of the *Great Soviet Encyclopedia* that appeared in the 1970s, Stalin rated only a page and a half (compared to thirty-one pages, for example, for the

Latvian Soviet Socialist Republic), a dry listing of dates, posts held, and a few stilted remarks like, "certain of his character traits had negative repercussions." Statisticians were not allowed to try to calculate whether these negative repercussions sent 20, or 25, or merely 18 million of their countrymen and women to early deaths by famine or by the executioner's bullet or in the *gulag*. Economists were not allowed to analyze the problems of transforming an economy where more than 20 percent of the national labor force had been prisoners. The Politburo member in charge of ideology told the novelist Vasily Grossman that his long novel about the Stalin years, *Life and Fate*, couldn't be published "for two or three hundred years." Just as we remember the late 1930s in Russia by the Great Purge, this later period—most of the 1960s, the 1970s, and the first half of the 1980s—we can call the Great Silence.

For roughly a quarter of a century, the entire wrenching, bloody Stalin era, which had devoured so many mothers, fathers, aunts, and uncles of people now alive, was officially unmentioned. People had ways of recognizing each other among strangers: if you said that your sister or grandfather had "lived for many years in the Far East," someone else who had lost a family member to the *gulag* would know what you meant. Otherwise, Russians only talked about the subject behind closed doors, or in the furtively circulated typescripts of *samizdat* literature. This constant, looming, unacknowledged catastrophe was, as therapists say of long-denied family secrets like incest or alcoholism, "the elephant in the living room" that no one dares speak of. And, like an individual survivor of prolonged abuse, a whole society that survives shows the same post-traumatic symptoms: Amnesia. Shame. Numbness. And a lingering fear.

In 1985, Mikhail Gorbachev took office and the first stirrings of *glasnost* began. In the exhilarating several years that followed, doubtless the first elections and the end of the Cold War mattered more, but I found that what moved me most was news of how Soviet writers, artists, filmmakers, and ordinary people were now finally starting to talk openly about that long-suppressed past. *Washington Post* correspondent David Remnick called it "the return of history." The Great Silence was finally broken. Russia was like a person who had endured unimaginably terrible suffering as a child, then for many years was strictly forbidden ever to mention it, and who now at last, in middle age, was able to speak.

The words came pouring out. By 1987, previously forbidden books about the Stalin era began to appear, and in 1988 they became a flood. At public meetings, victims of the Great Purge stood, wept, and told their stories of torture, prison, and exile. Plays that had been forbidden for decades filled the country's stages, and powerful new documentary films raised unsettling questions about how the country could have let itself be mesmerized by Stalin for so long. At an exhibit in Moscow about the *gulag,* visitors laid flowers in a labor camp wheelbarrow; on the snowy street outside, in midwinter, the line of people waiting to get in was more than half a mile long. From the country's Supreme Court and Politburo came announcement after announcement that famous revolutionaries shot after the show trials of the 1930s were now rehabilitated. The Politburo appointed a special subcommittee to study the repression of the Stalin era. It is hard to recall another case where the government and press of a major country were so preoccupied, almost daily, with events that had happened forty or fifty years before.

One part of the Soviet past seemed particularly to demand confronting. If the metaphorical center of the Holocaust is the railway platform at Auschwitz in the months when Dr. Mengele stood there making his famous "selections," the equivalent time for Stalin's Russia would be around the year 1937, at the height of the Great Purge, and the place would be Kolyma.

For Russians, Kolyma—pronounced Kol-ee-MA—is, like Auschwitz, a shorthand word, the final, fatal destination, the part that stands for the whole. With the local names changed, you hear the same joke in every city in Russia:

"What's the highest building in St. Petersburg?"

"It's St. Isaac's Cathedral."

"No it's not. It's the Shpalerka [the secret police interrogation prison]. Because from the top floor, you can see all the way to Kolyma."

To find Kolyma on the map, run your finger across the Bering Strait from Alaska, continue a few inches to the left and there you will find a swath of land sandwiched between the Arctic Ocean to the north, and, to the south, the Sea of Okhotsk—the arm of the Pacific above Japan, ringed with ice for several months of each year.

Part of Kolyma lies north of the Arctic Circle, and much of it is mountainous. The desolate, windswept Kolyma town of Oymyakon is the coldest inhabited place on earth. In the 1920s, the Soviets discovered that the region had a great wealth of minerals, including some of the world's greatest known deposits of gold. And so when millions of prisoners began flooding into the *gulag* in the 1930s, it was to Kolyma, to dig the gold from the perpetually frozen earth, that huge numbers of them were sent. Kolyma had well over a hundred known labor camps.

Even today, no railroad or year-round highway reaches Kolyma. Prisoners sent there in the 1930s, 1940s, and early 1950s traveled in boxcars or special prison cars thousands of miles across Russia on the Trans-Siberian Railroad. Next, they were held in one of the infamous "transit camps" at the Pacific end of the Trans-Siberian. Then came a week's journey in the hold of a freighter, northward across the Sea of Okhotsk, to Magadan, Kolyma's capital. Many prisoners died on the train. Others succumbed during the sea voyage. For those who still lived, the work now began.

Kolyma's gold-mining camps often had a death rate of 20 percent a year; many prisoners sent to the territory never returned. Some of these died in mining accidents; some were shot; most gradually starved. Russians spoke the very name of Kolyma in hushed tones. Then, as surviving prisoners were released after Stalin's death and came back to what they called "the mainland," they brought with them stories of this distant, wintry land of the dead. One of these survivors was Varlam Shalamov, whom Alexander Solzhenitsyn called "the finest writer of the *gulag*." Shalamov wrote poetry and spare, understated, razor-sharp short stories about this world of ice and death. In one story, a desperate escaper, guards and dogs closing in on him, takes refuge in a bear's lair. In another, a secret police interrogator reverently recalls the most thrilling moment of his career: the time he actually touched the file of Gumilev, the famous executed poet. In a third, two Kolyma prisoners slip out of their barracks at night to dig up a just-buried corpse; they want its clothing, to trade for bread: "Everything seemed real but different than in the daytime. It was as if the world had a second face, a face of darkness."

Another writer who survived Kolyma was Eugenia Ginzburg, author of two vivid, searing volumes of memoirs about her seventeen years of prison and exile. Like Shalamov, she saw her books published to great acclaim in the West, but did not live long enough to see them printed in Russia itself today. "Even the stranger whom you meet on your travels . . . immediately becomes near and dear to you when you learn that he was *there,*" wrote Ginzburg of the *gulag*. "In other words, he knows things that are beyond the comprehension of people who have not been there, even the most noble and kind-hearted among them."

As *glasnost* unfolded and I followed the new Soviet debate over the Stalin years, I wondered what life was like *there,* in the *gulag* territory now, especially in remote Kolyma, where skeletons in frozen, shallow mass graves far outnumber the living. There are so many bones still lying about, said one account I read, that today in the summer Kolyma children use human skulls to gather blueberries. What was it like to live and work daily amid such reminders of mass murder? Surely, for those who still live in Kolyma, dealing with the past would be an even more painful job than for people in the rest of the country. Until recently, it was impossible to find out: Kolyma was still closed to foreigners.

This restriction, however, was one of many that was lifted as the Soviet Union slowly began to crumble. This moment in the country's history felt like the opening of a great window. I had made half a dozen shorter visits to Russia as a journalist, spoke the language passably, and had always wanted to go back for a longer stretch of time. Now, at last, there was an opportunity I had thought might never arise in my lifetime: to see how Russians were breaking the Great Silence and dealing with Stalin's ghost.

With *glasnost,* it now seemed possible to find all sorts of people whom it would have been difficult or impossible to speak to before. It was less than four decades, for instance, since Stalin had died. There must be people still alive who worked with him. How did they feel about him now? The liberal weekly *Moscow News* was running stories about a notorious torturer for the secret police of the 1930s, whom a reporter had just discovered alive and well and working as a scientist in Moscow. This was a man who had once interrogated a woman for eight days straight, then ordered her hung up by her hair until she died. Of the thousands like him, could I find any willing to talk? Would they be repentant? Or defiant? Besides the executioners, I also wanted to talk to the victims. There were an estimated one hundred thousand people still alive who had been in labor camps during the Stalin years. How did they now look back on that time?

Then there were the mass graves turning up all over the various republics of the fast-dissolving Soviet Union. In earlier years, if somebody stumbled onto a pile of human bones, it had been immediately hushed up. But now the Soviet press was filled with news of these discoveries. Workers laying a gas pipeline through a pine forest near Minsk came upon one of the largest sites: an estimated one hundred thousand or more skeletons dating from 1937 to 1941, some still clutching the items they were holding when arrested—eyeglasses, medicines, purses full of coins. At some grave sites, corpses were lined up in pairs: executioners had saved bullets by killing two victims with each. People found hundreds of thousands more skeletons jammed into abandoned mining tunnels in the Urals, yielded up by the permafrost in Siberia, and beneath grass and trees in hidden corners of Moscow itself.

In most societies, barriers of time and space surround such history. The worst deeds on our own collective conscience—slavery, and the slaughter of the American Indians—are safely back in the last century and beyond. And, for more recent events, we usually have an ocean in between: no American child digging in a backyard is going to stumble onto skeletons of the Vietnamese who died at My Lai, or the victims of the U.S.-financed death squads in El Salvador. Yet people in Russia have no such barriers. What would it be like, I wondered, if beneath a familiar park or street you suddenly found a mass grave—which might contain the bones of your own relatives? I wanted to visit such a town and try to find out.

Finally, there was that larger set of questions, which Westerners have been debating for decades, but which Russian intellectuals have become free to talk about openly only now: about what allows people to become executioners, about the two faces of Utopianism, and about seeing and denial.

I planned a stay in Russia of some six months. My family and I arrived in Moscow on a snowy night in January 1991, the beginning, as it turned out, of Gorbachev's final year in power and of the Soviet Union's final year as one country. In the building where we found an apartment, we were the only foreigners. An immense structure dating from the Stalin years, it was built like a fortress around a central courtyard. The dirty stairwell was lit by what seemed a 10-watt bulb, and the small elevator sounded like a meat grinder. But double-paned windows and thick stone walls kept out the cold. The building stood between a factory still named after Felix Dzerzhinsky, the first chief of the Soviet secret police, and an attractive little park still named after Pavlik Morozov, the fourteen-year-old boy who became an official hero in the 1930s for denouncing his own father to the authorities. It seemed an appropriate spot to begin my journey of exploration back to that dark time, through the memories of people still living. Before the end of those travels, I wanted to talk with Russians of every sort to see how they were remembering, judging, and coming to terms with that period—and to talk not only with people in Moscow, but with men and women across this entire vast country, all the way to the icy heart of darkness itself, Kolyma.

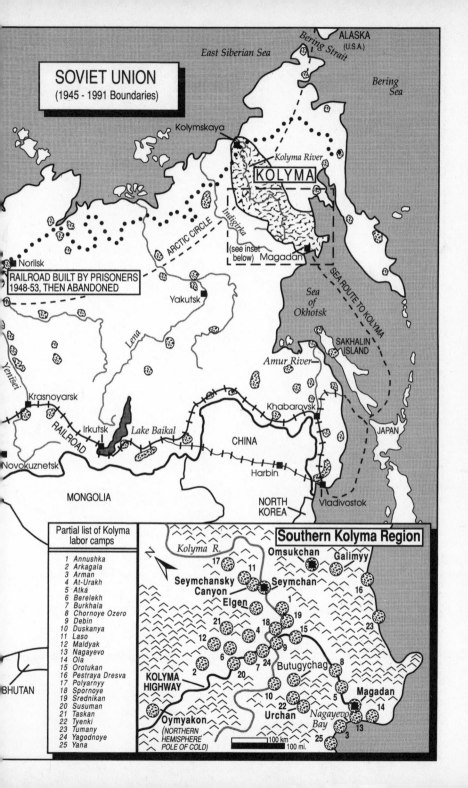

The Party organized thousands of meetings to unanimously demand death sentences for "enemies of the people" on trial. These workers are at a Leningrad machine shop in 1936.

1

Father
and Son

In the dark, grimy Moscow midwinter, we could see the sun for only half a dozen hours in the middle of the day. For one stretch of a week or so, the temperature went down to -35° F. at night. The cold seeped like liquid through the tiniest of openings; my wife and teenage son and I slid strips of newspaper into the cracks around our frost-rimmed windows, then taped them over. As I was out interviewing someone in an unheated room one day, the ink froze in my pen. I still wanted to go to Kolyma, but the record winter low temperature there is -97.8° F. To make that trip, I decided, I'd wait for warmer weather.

In the neighborhood around our apartment, long lines of worried-looking people in fur hats and heavy coats waited in front of grocery stores, their breath condensing in little clouds in the air. Inside the stores, the vegetables were wilted; the meager amount of meat on sale was all bone and gristle. The items that did fill the shelves were things nobody wanted, like immense jugs of vinegar. The frozen, snow-covered surface of the Moscow River was dotted with people fishing through holes in the ice; what had once been a sport was now more important as a source of food.

With so much anxiety in the air about the present, was anyone still thinking about the past? They were. As I explored the cultural landscape in these first few weeks, a surprising amount had to do with the Stalin period. In the auditorium of the Writers Union, several hundred people turned out for an evening of poetry and music by survivors of the Kolyma camps. The Sovremennik Theater was staging a play based on Eugenia Ginzburg's prison memoir, *Journey into the Whirlwind*; on the day I went, every seat was full, although the play had already been running for two years and theater attendance has dropped drastically with the economy. "Leave me in peace," begs Stalin's ghost in another play, by Mikhail Shatrov. A character replies: "If you only knew how little we want to have to talk about you. The problem is that whatever we turn to today, we find ourselves looking at you."

The subject of the past was so laden because it was also, symbolically, a discussion of which way the country should go now. The past was, as one journal put it, "tied by a thousand threads to the present." And so the debate still raged several years after it began, and my files soon bulged with material clipped from the newspapers daily. Each new revelation in the press brought a storm of letters to the editor, and in turn rippled through people's daily lives. "We enter into a kind of moral dialogue," wrote the reformist historian Yuri Afanasiev, "with those who lived in the '30s." This dialogue was often painfully personal. For it was not just a matter of facing what the country did forty or fifty years ago, but of facing what a father or mother or grandparent had done, and had long concealed.

Again and again I heard stories of how the recent digging into the Stalinist past had brought memories to life and shown them in a new and disturbing light. In one friend's family, an uncle had been "repressed," to use the Soviet euphemism, in the Great Purge. Now my friend asked her father, a long-time Party member, "What *did* happen to that uncle of yours? Why was he shot?"

Her father replied anxiously, "If anybody ever asks you, when you're applying for a job or anything like that, 'Was any member of your family repressed?,' answer *no.*" Suddenly she realized that he had been doing this all his life, living in fear of being found out.

A Moscow clinical psychologist told me about another case. A woman patient from years earlier had recently returned for more treatment, in deep distress. To her horror, newly published accounts told how her father, a diplomat, had denounced many people to the secret police and was responsible for their deaths. It was not by accident, she understood now, that he had remained alive and been promoted while his colleagues had perished. He had been dead for some years, but this woman now had to come to terms with her father all over again—something far more painful the second time.

If I still had any doubt that the wounds of the Stalin years could reach deep into the daily lives of Russians today, it vanished one morning a few weeks after our arrival in Moscow. A Russian friend had said on the phone, "There's somebody visiting town you really ought to talk to." He paused, and added, "Ask him about his father."

And now here the visitor was, sitting across our kitchen table as I put out food for breakfast: an erect, balding, fifty-five-year-old man with a graying beard and mustache and intense blue eyes that rarely left my face. His name was Nikolai Danilov.

Danilov is not a survivor of Stalin's *gulag,* and no one in his immediate family was killed by the dictator. Yet the events of those years have fixed the course of his entire life, and they determine it still. The story he told begins in Russia in 1938, as the shadow of mass terror hangs over the entire country. And it ends fifty years later, in a hospital room on the other side of the world. It is the tale of a family torn apart by fear, and of a restless, driven, searching man determined to put it together again. And contained within this story is an even more unusual one, of the searcher himself: a passionate human rights activist who was once an officer of the KGB.

In the second half of the 1930s, the paranoia of the Great Purge raged across the Soviet Union like an epidemic, targeting almost every imaginable group. Their members would fall victim, arrested and shot by the tens or hundreds of thousands, accused of conspiring against the government. Old Bolsheviks—Party members from before the Revolution. Scientists who would not subscribe to Lysenko's wishful-thinking biology. Engineers charged with sabotaging the factories where they worked. Anyone who had traveled abroad. Stamp collectors and Esperanto-speakers (suspicious foreign ties). World War I veterans who had been German or Austro-Hungarian prisoners of war (they might have been recruited as spies). One of the most bloodied groups of all was the Red Army's senior officer corps: 43,000 military officers were seized. Among them was Nikolai Danilov's father.

Most of these arrested officers disappeared into mass graves or the distant camps of the *gulag*. But for some unknown reason—the grounds for mercy in these cases were no more rational than the grounds for a death sentence—Danilov's father was lucky. After some time in the custody of the NKVD, as the secret police was known at that time, he was released and returned to army duty. He was back in uniform, living with his family in officers' housing, on the fateful day when some 3 million German soldiers stormed across the Soviet frontier.

"I was not yet seven years old, but I remember June 22, 1941, perfectly well, when the war started," said Danilov. "I was perched on the windowsill. It was a Sunday. I heard a loudspeaker outside: 'Today, at four o'clock in the morning, a treacherous attack ...' The speech ended with the words, 'Our cause is just and victory will be ours.' My mother, my sister and I had just had breakfast, and my mother was doing the dishes. Suddenly a plate dropped from Mother's hands and broke into pieces."

Danilov's father was a lieutenant colonel. "The rank of lieutenant colonel was introduced on the eve of the war; it was a very rare rank then. He was sent to the front. I remember my mother seeing him off. She made the sign of the cross over him."

Soon bad news came from the front, where Soviet soldiers were suffering terrible casualties at the hands of the invading Germans. Danilov's father was reported missing, presumed dead.

Toward the end of the war, when the Germans had been

pushed back, Danilov and his mother and sister moved to the newly liberated city of Kremenchug, in the Ukraine, where Danilov's grandmother lived. There Danilov grew up—the child of one of the many millions of Soviet soldiers who never came home from the war. At church, Danilov noticed, his mother still lit candles for his father's health. But he dismissed this as a widow's false hopes.

Three years after Stalin's death, Nikolai Danilov finished his education. It was 1956, the year that Khrushchev first denounced Stalin in his "secret speech." Some 7 or 8 million prisoners were released from the *gulag:* "From the bottom of the sea," one poet wrote, "Friends began to return." The promise of change was in the air; no longer did everyone live in fear of the midnight knock on the door. The dreaded Lavrenti Beria, Stalin's last NKVD chief, had been deposed and shot. And the secret police had recently changed its name, to the KGB. Looking for a job, Danilov applied to join it.

One of Danilov's main qualifications, it seems, was that he was not Jewish. A school friend who was heading for a career in the agency had suggested that Danilov do the same. "Why don't you come to the KGB?" his friend said. "They told us that if we knew people with a higher education—and they stressed that these guys should not be Jews—we should send them along. You know, things are different now. Beria has been executed. The job is very exciting."

"And he gives me," Danilov said, "the telephone number of the personnel department, and I walk into the KGB. I'm thinking that it's a time when justice will be restored, and that I'll be participating. I decide to take the risk. They give me a thick questionnaire to fill in. Of course I don't mention that any family members were repressed. As far as my father is concerned, he is known to be lost in action in the war.

"My mother said, 'Kolya, what are you getting yourself in for?'"

Danilov spent a year in KGB training. In a secluded building, the new cadets studied "eavesdropping, surveillance, how to recruit an informer, how to start working with him, how to keep records, possible places for clandestine meetings, secret addresses, all that stuff known the world over."

Then Danilov was assigned to work on rehabilitations. At this time, the late 1950s, the government was legally rehabilitating thousands of victims of the Great Purge, both the living and the dead. Each case had to be laboriously examined by a special investigator like Danilov.

Danilov did some of his rehabilitating in the town where he had grown up, Kremenchug. During this work, he noticed that the KGB was burning huge numbers of old records. He wrote a poem about these burned archives at the time, and he recited it now, across the breakfast table, in a deep, melodious voice full of rolling r's:

> They have seditious ideas, forbidden thoughts;
> They hide secrets . . .
> Flames reach them with copper-red fingers.
> . . . betrayal is writhing, cowardice is squirming in the flames
> The papers are crying out in bursts of fire . . .

Immersed in documents about mass arrests, interrogations, and forced confessions, Danilov began to ask himself questions. "I tried to understand who had been to blame for all this. I began to realize that Stalin alone could not be. Who was to blame? The interrogator? The interrogator got all his 'evidence' from an agent in the field. The agent, though, had gotten a denunciation from somebody who claimed that somebody else was an enemy. Who was to blame? Gradually I began to understand that the system that encouraged denunciations was to blame. If there was a mailbox for denunciations, it filled right up. If there had not been that mailbox, denunciations might have been very rare."

Danilov soon ran afoul of his KGB superiors. "There was a Municipal Students Club. At a café for young people, they recited poems in the evening. Once a KGB colonel summons me, and he tells me, 'What are you doing with these students? There are lots of Jews there! Our agents say you're going there, but you haven't reported to us about what's going on there.'

" 'Why?' I say. 'I'm not an agent of yours. I'm a rehabilitations investigator.' "

The KGB did not like this kind of back talk, and in 1960, the agency transferred the uncooperative Danilov as far away as it could send him, to the Pacific island of Sakhalin. In this new post, too, he swiftly got into trouble, for complaining that the KGB was opening citizens' mail. Three years later, as a lieutenant, Danilov resigned from the secret police. Explaining his decision now, he quoted an old Russian proverb: "If you live with the wolves, you must howl with the wolves."

Danilov returned to European Russia, studied literature, and became active in the dissident movement. In 1968, as a member of a Leningrad human rights group that protested the Soviet bloc invasion of Czechoslovakia, he was arrested. The KGB also charged him with "revealing the methods of state security"—because of his poem about the burning of the archives.

He was put in prison, and then sent to a psychiatric hospital where they gave him insulin shocks. Danilov's wife left him while he was in jail. But he learned, he said, "that human values have supremacy over state ones. From the perspective of the state, I had betrayed it. But I didn't betray myself. That was the most important thing."

With his youth, his KGB period, and several years in custody behind him, Danilov moved to Odessa, on the Black Sea. He remarried and had a daughter, Anya. Then, in the early 1970s, he found out something that astonished him.

His father had not been killed in the war, after all.

Danilov's father, like so many Russian soldiers, his mother now revealed to him for the first time, had been caught behind the advancing German lines. Danilov's grandmother, the one who lived in Kremenchug, had told his mother that "during the Nazi occupation, Father, in disguise, in plainclothes, dirty and bearded, arrived in September or October of 1941, and stayed there until the very retreat of the Nazis." For nearly thirty years, terrified of the potential disgrace and danger it might cause for them all, Danilov's mother had not dared pass this secret on to her children. The Great Silence of the post-Stalin years was not just a matter of topics ignored in the newspapers or in school classrooms; it slipped deep inside many a family.

Stunned at this news, Danilov began to search for his father. He traveled to various cities to question war veterans. He looked at records in military archives. He wrote a magazine article under the headline "Honor Thy Father," and urged readers with leads to write him.

Slowly, he uncovered the story. His father's Red Army division had been surrounded by the Germans in one of the huge, disastrous pincer movements in the opening months of the war. Those soldiers who could not break out of the German encirclement were ordered to go underground. That was when his father had lived in Nazi-occupied Kremenchug.

And, Danilov discovered, in Kremenchug during the war there was an underground group of anti-Nazi partisans, headed by a "Lieutenant Colonel Petrov."

"I told you that lieutenant colonel was a very rare rank before the war." Other evidence also suggested that this "Petrov" was his father. And Danilov also discovered that a woman who had sheltered "Lieutenant Colonel Petrov" in Kremenchug had, like him, disappeared from the city toward the end of the war.

To a Westerner, it might be hard to understand why a Soviet partisan leader would not gladly welcome the liberation of his city by the Red Army. But Danilov's father knew he had two strikes against him.

First, he had already been arrested once as an "enemy of the people" before the war. People who had seen the inside of NKVD cells in the 1930s lived in constant fear that at the slightest false step they could fall back into the hands of the secret police. Second, as a Soviet officer, Danilov's father would have been taught that the worst possible crime for a Red Army commander was to let his unit be surrounded and his soldiers captured. In the early days of the war, when the Germans on several occasions captured hundreds of thousands of Soviet troops at a time, most of these men had no control over what was happening to them. But that made no difference. When the war was over, almost all Red Army soldiers who survived Nazi prisoner-of-war camps were sent straight to the Soviet *gulag*.

Even though Danilov's father had not been taken prisoner, many partisans feared—often correctly—being treated equally harshly when the war was over. Knowing that his father must have made these calculations, Danilov figured he had fled west. But where to? Danilov had one clue, misleading, as it turned out later. Word reached Russia that a man who had worked with Danilov's father in Kremenchug during the war had died in Philadelphia.

Only the most privileged Soviets could travel abroad. For someone with Nikolai Danilov's past as an imprisoned dissident, there was no hope of getting a passport. Instead, he pressed everyone he could find to ask their relatives abroad to look for traces of his father. He wrote letters to Europe and America. A Russian-language newspaper in New York published the story of his search.

Finally in 1981, some ten years after Danilov began his quest, word came back. A Nikolai Semyonovich Danilov had been found—in the telephone book for Winnipeg, Manitoba.

From Odessa, Nikolai Danilov, the son, wrote a letter.

There was no answer.

"When I told Mother, she said: 'And what if it's not him, what if it's somebody else with the same name?' "

Danilov tried again, this time mailing a registered letter to Winnipeg. "My father didn't realize that they would return the receipt he signed. And that I could compare this signature with a 1941 envelope. His D is very distinctive, Gothic"—he demonstrated with his finger on a table napkin—"then the A is very large, and then the letters become smaller and smaller." Even though the signature on the registered mail receipt was in the Roman alphabet and the samples of his father's handwriting from forty years earlier were in Cyrillic, the resemblance was unmistakable. It was the same man.

"So I telephoned him. I had prepared myself for this talk. Nevertheless, I was anxious. I stumbled. He was as calm as if we had parted only yesterday. 'Ah, is that you, Kolya?'

"I said, 'I want to see you, so much.'

"He said, 'I'm old and sick, I'm not able to come.'

"I said, 'You can invite me to come there.'

"He said, *'They'll never let you out.'* "

For the better part of a decade, this remained true. That first telephone call was before Gorbachev. Then, with *glasnost,* Soviet restrictions on traveling abroad began to loosen. Nikolai Danilov was determined to visit his father, but the government still would not give him a passport. He wrote to the Interior Ministry. He wrote to the Supreme Soviet. He wrote to Foreign Minister Shevardnadze.

Only after he wrote to Gorbachev did anyone tell Danilov the reason his application for a passport was denied: "I knew state secrets! Then I demanded that they tell me what kind of secrets I knew. I had taken off my KGB shoulder boards more than twenty years before! Since then I had not been exposed to any secrets." Danilov wrote to the U.S. Embassy. He wrote to the secretariat set up to supervise the Helsinki Accords. He wrote to an American diplomat negotiating human rights issues with the Russians. Finally, in 1988, he got his passport.

For more than an hour now, there had been food on the table in front of Danilov, but he had not touched it. He was a man aching to tell his story; his eyes burned with it; his life revolved around it; he had poured his entire being into this quest—perhaps even his KGB period had been an unconscious preparation for it. He would tell his story again and again for the rest of his life, it seemed, and it would never be enough.

"I spoke a couple of times on the phone with my father," he continued, "and he kept saying that they would never let me go." In this insistence, Danilov began to read clear signals that his father did not want him to visit Canada. His father had married again, he discovered, to the woman who had sheltered him in Kremenchug during the war. They had fled to the West together, and no doubt dreaded the appearance of a child from an earlier marriage. The son would be a reminder of the family his father had left behind, in the nightmare conditions of war and fear of death, forty-five years before.

In any other society, a father who disappeared, abandoned his wife and children, and went off to a foreign country with another woman would be condemned. But if Nikolai Danilov's father had returned from the war, he would almost certainly have been sent to the *gulag*. By fleeing to the West, he probably saved his life. Moreover, by abandoning his family he helped them as well. Being the widow and children of an officer presumed killed in action was an honor, with material benefits too. But being the family of an "enemy of the people" carried a great stigma, and sometimes meant prison for the wife and an orphanage for the children. And so Nikolai Danilov did not feel he had anything to forgive his father for. He only wanted to see him again, before the old man—who had already spent some time in the hospital—died.

Sensitive to what he guessed were his father's conflicting feelings, in 1988 Danilov finally made the long journey from Odessa to Winnipeg. He took his thirteen-year-old daughter Anya with him. Wanting to figure out the least threatening way of approach, Danilov went first to the hospital, to try to find a doctor who could give him news of his father's mental and physical condition.

To Danilov's surprise, he learned that his father was back in the hospital again. The receptionist told him the room number.

"I open the door, and I see him lying there with his eyes closed. He and I look very much alike. If I had my beard shaved, we would look the same. He lies there and we stand there paralyzed. Anya says that she recognizes him from his pictures. He opens his eyes, props himself up in bed, and we come in.

"I say: 'Hello Papa. Do you recognize me?'

"He says, 'Yes. How did you come?'

"His bed goes up and down and we fix it so he can sit. I take one hand, Anya the other. His hands are cold, then they become warmer. He holds my hand and he rubs and rubs my nail with his thumb. We talk and I tell him that his mother died, and about his sisters, and about everything. The only thing I do not tell him anything about is my mother. All that was for him as if from another life. I think: if he asks, then I will tell him.

"We speak for two and a half hours, and we plan to come again the next day. Ekaterina Ivanovna [the second wife] will be there and we will meet her. I will come every day. He is in a wheelchair—I will wheel him around. And of course I still have many questions, such as about whether he had been this 'Lieutenant Colonel Petrov.'

"The next day we came. The bed was empty.

"I asked the nurse, and she told us that his wife had come and taken him home.

"When I phoned them, it was she who answered. 'He said that either he'd had a dream or that his son had really come to him. So as not to upset him, I told him that it was a dream.'

"I said, 'Do you really think that I've come here, so many thousands of miles, to have you tell him this is all a dream?'

"Then she put him on the phone, and he began to talk, but not the way we had talked the day before. He said, 'I didn't expect my son to come here to kill me!'

"I said, 'Papa, who is killing you?'

"He said, 'Yes, yes, to prove to the NKVD that you've got me!'

"Can you imagine? He'd been living all this time expecting the NKVD to come and arrest him, and the first person who reached him was me. So all these fears were focused on me. He had lived so many years, in a free country—and had still not gotten rid of such fears. Forty years had passed! Can you imagine what kind of fears have been sown in *this* country if a person could maintain them through his entire life over there?"

Danilov's blue eyes still showed the same intensity of feeling. What was it like, I asked, when he next saw his father?

"We did not meet again.

"I went to friends of his, who wanted to arrange a meeting for us. They told me that they had invited him and her for a certain time. But that she, Ekaterina Ivanovna, said that if my daughter and I came, they would not come."

Ekaterina Ivanovna was worried, Danilov figured, that he would make some kind of claim on their property, as his father's heir. But that was not what he wanted. And, however irrational her fears or his father's, Danilov did not want to force himself on them.

"I did everything possible to arrange another meeting. I could not have done more. I stayed on in Winnipeg for a while. I went to the Russian church, met all the Russian émigrés. A lot of them knew him; they all invited me over. I would visit one family one day, another the next. They gave me pictures of Father. I have a whole pile of pictures, starting with the first ones in 1949 [when he arrived in Canada]—young, joyful, the war has just ended and he has escaped from the USSR. Wonderful pictures. The first ones were black-and-whites, then they were in color.

"Before my departure I wrote a letter to my father to say that he should not worry, that I was leaving, that it was a pity the way things had happened. But that I had always known that he was my father.

"A year and a half passed. On November 24, 1989, he died. All the answers to my questions he carried to the grave. . . ."

2
The
Uses of
Memory

Life is often more complicated than the lessons we try to draw from it, and I am wary of diminishing Nikolai Danilov's story by too quickly claiming a moral for it. But his long quest does suggest the sheer weight of the Stalinist heritage: in this family, it had broken the most basic of human bonds, between parent and child. One reason that era's legacy lies on today's Russians so heavily, I suspect, is that human beings have a longer memory for fear than they do for most other emotions. This is especially so when the fear comes not from something temporary, like a sudden illness or a natural disaster, but lasts for decades and is woven into a society's very framework. For almost all of Soviet rule, informers were everywhere; the NKVD had offices not only for every town and city, but in railroad stations, hotels, universities, and every large factory. Figuratively, and often literally, there was a "mailbox for denunciations," as Danilov put it, in every Soviet town and workplace. And even in every home: "Nobody confided their doubts to their children," writes the late Nadezhda Mandelstam, one of the most eloquent witnesses of the time, ". . . suppose the child talked in school and brought disaster to the whole family?"

The memories Nikolai Danilov carried with him seemed like a scar that would never fully heal. Yet what I found most impressive about him was that today he spends all his spare time helping others deal with their own scars from the past. He has become the leading researcher on mass graves in Odessa, where he now lives, writing articles on the subject, excavating grave sites, looking for clues to which groups of people were shot when. He has started an informal network that helps other people find family members missing since Stalin's day, both the living and the dead. And in this work of healing and remembering and overcoming the Great Silence, he is not alone, but is a member of an unusual national organization.

Evening: a one-room office, heated, mercifully, against the subzero weather. The office opens off an obscure back corridor on the ground floor of an apartment complex on the northern outskirts of Moscow. The room is dimly lit. It is barely big enough for two desks and two chairs. On the walls are posters of Andrei Sakharov, addresses and phone numbers of government offices, and open envelopes marked with people's names, tacked up, back side out, to serve as mailboxes. The homey clutter of the bulletin board has the look of low-budget grass-roots organizations anywhere in the world. Here are two of the notices posted on it:

• Anyone who knows Sergei Timofeyevich Dobrovolsky (former school principal from Donbas, repressed in 1930), or who was in prison with him, or who has heard anything of his fate, please call Moscow Memorial at 435-0475 or write to: Vladimir Sergeyevich Dobrovolsky, Apt. 36, 16A Lenin St., Ilichevsk.

• In 1937–38 all our family—six people—were arrested, including three children: my daughter Tamara (born 1934); and my niece and nephew, the children of my sister who died in 1934, whom I was taking care of—Lila (born 1927) and Yuri (born 1931). After the entire family was arrested, the children were sent to various orphanages. I was five years in prison starting in 1938. . . . After 10 years I found my daughter, who was in an orphanage in the Urals. In 1971 I found my niece. To this day I haven't found my nephew. I ask Memorial to help me find my nephew, Yuri Petrovich Antonov (born 1931). Thank you . . .

The office is a neighborhood outpost of Memorial. This group (pronounced Memori-AL) was founded in 1987, and today has more than one hundred branches in Russia and the other republics. Memorial is loose, informal, non-hierarchical; its members include labor camp survivors, those who lost relatives to the camps or the firing squads, plus writers, historians, and human rights supporters who want to lay bare the history of the *gulag,* to remember and help those who suffered, and to make sure it doesn't all happen again.

Powerfully influenced at its beginnings by Andrei Sakharov, Memorial's call has always been for truth—above all, for the complete opening of all official archives—and not for vengeance. How could Russia ever stage Nuremberg Trials, one Memorial member who lost a grandfather to the Purge said to me, when those complicit in the *gulag* included "practically the whole country— one part denouncing, one part judging, a third shooting people, a fourth guarding the camps. It wouldn't stand to reason to try to find out who are more guilty than the others and to prosecute them. It would trigger the same conflict all over again, but in the opposite direction."

"But," adds a journalist writing in *Moscow News* about former secret police officials, "give their names—we must. We must name them because these people lived and live among us."

Memorial systematically gathers information from former prisoners and their families, and has recently begun transferring this data onto computers. One reason for doing this is so copies can be stored in many places: right now the authorities are not harassing Memorial, but in Russia you never know. The other advantage of computerization is that it allows cross-indexing: a son or daughter can find someone who might have shared a barracks with a missing father; a researcher can trace patterns: were all the biologists, the historians, or the Party officials in a particular city arrested at the same time?

Memorial is also part relief and welfare organization for aging veterans of the *gulag* who are in need. Interestingly, one thing Memorial does not do is to proselytize aggressively. Many members will gladly talk about their experiences if asked, but the organization deliberately has no program to send speakers out to schools or elsewhere. In a society numbed by propaganda for seventy years, Memorial members are wary of thrusting their message at anyone who may not want to hear it. Younger members remember from school the scores of patriotic talks by aging World War II veterans. Russians have a wonderfully contemptuous phrase for such people: "wedding generals." "The kids give him flowers and forget him right away," one woman in Memorial who survived several labor camps told me. "I don't want them to make vets of us."

The core of Memorial's work is to try to restore a set of memories the government worked for several decades to erase. "We have to get over our loss of memory," writes Nadezhda Mandelstam, whose husband, the poet Osip Mandelstam, died as a prisoner on his way to Kolyma. "This is the first task . . . otherwise there will be no way ahead."

One major use of memory is simply to record and honor those who suffered or died. It is hard to easily encompass everyone: Nikolai Danilov's father, who still lived in fear of the NKVD half a world away, falls into no official category of victim. But for those who do, Memorial does what it can. One volunteer I met, herself a camp survivor in her eighties, was tape-recording interviews with other *gulag* veterans. Another had put together a traveling exhibit of paintings and drawings by prisoners; these were often done in secret, using feathers, sticks, or matches for brushes, and paint stolen from sign-painting workshops. Another way Memorial tries to honor the victims is simply by printing, to the extent they can be recovered, the names of the dead. Again and again, all across Russia, people showed me lists the local Memorial chapter had published for their city or district, in paperback volumes like small telephone books: thousands upon thousands of names, with dates of birth and death, professions or occupations.

In many cities, under pressure from Memorial and other

groups, the KGB has also released photographs of Purge victims. Wherever this has happened, a newspaper usually publishes a column of these pictures every day, gradually, over the months and years, working its way through thousands of victims. The paper in the capital that does this, *Evening Moscow*, runs a daily, black-bordered column with photos and thumbnail sketches of nine or ten secret police victims, almost all shot in the late 1930s. It is eerie to see these faces, a number clipped to each person's chest, staring out of the paper every day of the week. They do not wear the sullen frowns of convicts in American police mug shots, but have the shocked, numbed expressions of men or women arrested for the first and last time, knowing that they may die.

Memorial is an embodiment of a promise that many survivors once made to each other. One prisoner's memoir records a message scrawled in a camp latrine: "May he be damned who, after regaining freedom, remains silent." In an often-quoted passage, the poet Anna Akhmatova writes:

I spent seventeen months in lines outside Leningrad prisons. One day someone recognized me. Then a woman standing behind me with blue lips ... woke from the stupor in which we were sunk, and spoke into my ear (for there we spoke only in whispers):

"And can you describe all this?"

"I can," I said.

Then something like a smile slipped over what had once been her face.

Memory made public is also a warning: *you can't get away with this again*. In writing about the recovery from authoritarian rule in Brazil and Uruguay, Lawrence Weschler comments, "Retrospectively, the broadcasting of truth ... upends the torturer's boastful claim that no one will ever know. Prospectively the broadcasting of truth ... is at once more subtle and perhaps more momentous. ... It is essential to the structure of torture that it take place in secret, in the dark, beyond considerations of shame and account. ... Torturers can never again feel so self-assured—nor their victims so utterly forlorn."

Václav Havel speaks of the "fear of history" that leads people to avoid dealing with complicity and guilt. But sometimes we blot out the past simply to spare ourselves unnecessary pain. Often those who have made memory their business have evolved rituals to help them. On the anniversary of his arrest each year, Alexander Solzhenitsyn eats nothing but the *gulag* ration of 650 grams of bread and a few sugar cubes in hot water. Primo Levi, unlike many other survivors of Auschwitz, did not have the tattoo of his identification number removed from his forearm. He wanted people to see it, and to know or ask what it was.

Lev Razgon embodies this same spirit. At eighty-three, he looks barely seventy, spry and fit in a blue windbreaker. His warm, expressive face has blue eyes that fix you, long and thoughtfully, as if he is quizzically turning you over in his mind. He seems at peace with himself, curious about everything in the world around him and bemused that he is still alive to see it all. Razgon, who is on the board of directors of Memorial, is a writer and editor of children's books. He spent more than fifteen years in labor camps, most of it felling timber near the Arctic Circle.

I talked to Razgon in the famous "writers' village" of Peredelkino, some forty-five minutes' drive outside Moscow. It was here that Boris Pasternak lived and worked, and many other writers live here still. Razgon was staying in an upstairs room of the Writers Union guest house, whose window looked out onto the upper branches of a grove of birch trees.

"I started school before the Revolution," said Razgon. "In all our classes we would sing, 'God Save the Tsar.'" As a young man he married the daughter of a high Communist Party official, who was purged by Stalin. The NKVD then arrested Razgon, his wife, and many of their relatives. "Stalin destroyed not only his potential opponents but all their families. In this he followed Ivan the Terrible's example. He arrested wives, children. . . . Only my daughter was not arrested, because she was fifteen months old." Razgon's wife died in a prison train, at the age of twenty-two.

"I still regard March 5 [the day Stalin died] as a holiday. Beginning with the first day. I even managed to get vodka in camp. We chipped in: it was two hundred rubles plus ten cans of meat, and for that, a guard brought us a bottle of vodka. . . ."

More than twenty years ago, Razgon began writing his memoirs. At the time, he could only write "for the desk drawer, without any hope of it being published. Sometimes I read it aloud to my family and friends." But several years ago, his book was printed at last. It won several prizes; its translation into French brought him the first chance in his life to go abroad. At the age of eighty, he suddenly found himself a literary celebrity.

Razgon met a remarkable variety of people in camps and prison, and his memoirs are a portrait gallery of them all. One of his cellmates was an old man who had been in the very same Moscow prison cell under the last Tsar. One day the guards brought this man back from interrogation, badly beaten and unconscious. They dumped him in the cell, then fetched the prison doctor, a woman:

"We could not take our eyes off her. Not bending down, the beautiful woman in the white coat, using the toe of her small elegant slipper, pushed the hand of the man lying on the floor, and then his arms, which were stretched out on the asphalt floor in the shape of a cross, and then his legs. Then she turned to the warders and said, 'Nothing broken, just bruises.'

"She turned round and, glancing at us but not seeing us, she went out of the room."

At that moment, Razgon writes, he understood of such people that "They are not like we are . . . or like we will ever be."

The memorable encounters in Razgon's life did not end when he left prison. Once, some years later, heart trouble landed him in the hospital. He had the feeling that there was something strange about a fellow patient, Grigory Niazov. Razgon writes: "His eyes registered everything at once. His hearing was so subtle he could discern separate conversations in the hubbub in the hall. . . . He was always accusing [the doctors] of doing something wrong." Razgon began asking questions. Niazov, he learned, had worked at an NKVD "special facility" in eastern Siberia. He was a retired executioner.

Riveted, Razgon grilled him. "I memorized every word of Niazov's answers to my questions and will never forget them." The "special facility," Niazov said, had no purpose besides killing; prisoners were never kept there more than two or three days before being shot.

"We took them about twelve kilometers, to a small hill.... We'd shout: 'Get out! Line up!' They'd climb out, and in front of them there was already a pit dug for them. They'd climb out, huddle.... We shot them, and if anyone was still moving, we would finish them off and get back into the trucks.

"We would go back to the camp, leave our weapons in the guardroom, and drink as much as we wanted, for free.... I always drank one glass, went to the cafeteria, ate something hot, and then went back to the barracks to sleep."

Razgon asked Niazov: had he been married at the time?

"No, they didn't take married men."

Could he sleep after shooting people?

"I slept well, and during the day I'd hike around, there are some beautiful spots there."

How did he feel about the job?

"Of course it was a little boring, since there weren't any broads.... My conscience? No, it didn't bother me."

While interviewing someone who could have been his executioner, or in prison itself, Razgon never let his horror overcome his curiosity. "Camp and prison are great universities," he said as the wind swayed the birch branches outside his window. "Neither camp nor prison can make a bad person good, or a fool clever. But for a thinking person, it's a turning point in his life. I view the years I was there as the most important years of my life. Because camp was the only place in this country where a person could think freely and talk freely. In camp, people who tried to unravel what was going on had the scales fall from their eyes. The social layer that was arrested in 1937 and 1938 were highly cultured people. In free life one would have difficulty meeting people of such intellectual caliber. Of course many of them lost their humanity, because of cold, hunger, and humiliation. But many refused to break down. Some went through tremendous intellectual and moral tempering.

"Some try not to remember it, even to avoid other people who were in camp. They would like to cross out that period from their lives, to forget it. But others, including myself . . . I always remember it. Always. Every day. Every hour. Every minute."

Part of Memorial's mission is to heal, and one of the people I liked best in the organization was a doctor. Her work is yet another use of memory, perhaps the most important.

Marina Linnik is a cardiologist. A warm, compassionate woman in her late thirties, she had recently had a baby, and was on maternity leave from the intensive care unit where she worked. She was spending her time volunteering for Memorial, and with her family at home, where I visited her.

Had anyone in her own family suffered under Stalin?

Only a great-uncle, she replied, a prominent scientist. When he "came back" after Stalin's death, he was given an apartment and a summer dacha. The dacha was in a village where there were many other scientists who had returned from the *gulag*. As a child, Linnik walked along the road on summer evenings; by kerosene lamplight on the porches of the dachas, people would be reading Solzhenitsyn.

Linnik considers herself one of "the children of the sixties. We witnessed people coming back from the camps. Nobody would name it. It was just whispered: 'From *there*.'

"We saw the beginnings of *samizdat*. I remember so well those manuscript carbons on thin paper, like cigarette paper. How they rustled! My friends were in the so-called inner immigration. We kept a critical distance from what was said officially, from what they taught us in school. We were not the kind of people who cheerfully shouted 'Hurrah!' "

In this and other ways, Linnik is like a whole generation of Memorial activists. Although the group has its elders like Lev Razgon, many members I talked with or saw at meetings were in their thirties or early forties. Most were medical researchers, mathematicians, or scientists: in recent decades, Soviet science was free of the dogmas that clogged other fields and that is where many of the best minds went. Too young to have known the fear that so paralyzed their parents' generation, these "children of the sixties" found their curiosity lit by the jarring contradictions between the history of glorious Soviet achievements they studied in school and the very different history they heard around the family dinner table.

In the 1970s and 1980s, the "inner immigration" Linnik spoke of became, for Soviet Jews like her, an outer one. Today, most of the other children Linnik played with in that dacha village of her childhood are in the United States or Israel. From her circle of friends at medical school—she gestured at framed pictures on her bookshelf—more now live in New York than in Moscow.

Marina Linnik is one of a group of doctors in Memorial who volunteer their services to older members in need—usually *gulag* veterans or their children. At one point in our conversation the phone rang; she picked it up, listened, asked a question or two, and said, "Sounds like the flu. Why don't you call me again this evening and tell me how things are."

Along with the medical texts on the shelves of her study were many volumes of German literature. Linnik has visited Germany, and feels the West Germans have done a better job than the Russians of dealing with their past. The Germans have, she thinks, "a feeling of guilt, a feeling of responsibility, a kind of feeling which is not so common in this country. The Germans identify themselves even with the bad things. Yet this is not so here." In West Germany, Linnik said, she always sensed that it made people her age feel guilty when she mentioned she was Jewish. "But here nobody feels responsible for what happened under Stalin. Not at all. Even intellectuals here think in terms of *they. They* intervened in Afghanistan, *they* intervened in Czechoslovakia."

In the hallway of Linnik's apartment, along with a homey lit-
ter of crib, playpen, and children's toys, sat a large sack filled with
hundreds of letters. Among other things, Linnik now supervises
the answering of all mail that arrives at Memorial headquarters.
"Any letter that comes in?" I asked.

"Unfortunately, yes! All of them. What kind of letters are
they? They have information on people who were repressed. It
might be a response to our questionnaire, or stories about relatives,
or memoirs. We've sometimes had three thousand letters a week.
After several years now this stream has begun to lessen. But if a
newspaper publishes an article about us, it means a wave of mail."

Occasionally the mail is from someone "on the other side"—a
retired secret policeman whose conscience is bothering him. Or
whose conscience is *not* bothering him. "We had one letter saying,
'I've got inflammation of the nerves now, I've got pains in my legs.
I spent my life out there in Magadan guarding those enemies of the
people, and now you're rehabilitating them all, you bastards!'"

The vast majority of letters, Linnik said, are from "people who
had always concealed that their parents were 'enemies of the people,' were shot, were in labor camps. All their lives. Many did not
even tell their children. Then suddenly it has become possible to
speak about all this. It has been a tidal wave of information, all at
once. Or, say, someone is sorting the papers of an old man who just
died, and finds his journal; he had been silent for years and had
said nothing to anybody, not even to his children.

"Every letter gets an answer. It's not me who answers them,
but other volunteers, mostly older women who are themselves wid-
ows or daughters of the repressed. First a person sends in a letter
about her father, say. Then we ask her to go into detail about him.
Then she sends pictures. But by the tenth letter, everything is said
about the father, and then personal correspondence starts. And
that's good! Because in this way links are born, between people
who have spent all their lives alone with their misfortunes.

"Since I'm a doctor, I try to build these links as if they were
doctor-patient ties. I advise [the Memorial volunteer] to ask her
correspondent to write the second letter not simply to Memorial,

but personally to her, to the woman who answered the first letter. Many of these volunteers are lonely: their parents were repressed; their children are now grown up. All their lives they have lived with this tragedy. And now they begin corresponding with twenty people with similar fates. Some tell me that their correspondents have come to visit them, or that someone spent a summer vacation with a person she was writing to. Human bonds are established. And I'm very pleased. I think this is maybe more important than collecting material for the archives. . . ."

3

To See
Things As
They Are

Safely insulated by age or geography and looking at a period of great tyranny, most of us secretly think: *I* would have fought back boldly were I alive then. This tempting fantasy of heroism, however, makes a key assumption: the existence of supporters. The French Resistance fighter takes refuge in a friendly farmer's hayloft. The Abolitionist guiding slaves to freedom has a chain of safe houses. Or sometimes the support is potential, and the act of heroism awakens a nation's conscience: Gandhi's great marches and strikes in India; the sit-ins at lunch counters in the American South. But no such reservoir of support existed in Stalin's Russia. And few people even saw the regime for what it was, much less tried to overthrow it. These were the days when thousands of schoolchildren happily carried the dictator's portrait in May Day parades, and when cities all over the country were changing their names to Stalingrad, Stalinsk, Stalino, Stalinbad, Stalinir, Stalinkan.

". . . it is very difficult to look life in the face," writes Nadezhda Mandelstam. "To see things as they are demands a superhuman effort. . . . Many of [my friends] had awaited the Revolution all their lives, but at the sight of what it meant in terms of everyday life, they were horrified and looked away." Perhaps we should not be too hard on the Russians who looked away, because so many Westerners did, too. That steady procession of early American and European visitors to the Soviet Union mostly failed to see things as they were. (One delegation of prominent Americans even toured Kolyma, where nearly everyone was guard or prisoner, and found nothing amiss. I will come back to them later.)

In the Soviet Union of Stalin's day, that handful who did not look away needed roots in a moral universe of their own, for there was no community of fellow skeptics or rebels to support them. The small band of clear-seers, like Nadezhda Mandelstam and her doomed husband, were often men and women who came of age before the Revolution, with values rooted in the culture of Western Europe and of nineteenth-century Russia. Other brave people came from the rapidly dwindling band of oppositionist Bolsheviks, almost all in prison by the mid-1930s, who believed in the more idealistic and democratic part of the Russian revolutionary tradition. Among those who saw things as they were it was far rarer to find people born under Soviet power, who had never known another place or another time, and who were raised from birth in families and schools that questioned nothing.

And yet there were a few such people. One of them lives quietly in Moscow today. Her story is little known—and, until recently, at least one man who did know it had not even realized that she had survived.

One day in 1990, Miron Markovich Atlis, head of the Memorial chapter in Magadan, the major city of Kolyma, telephoned the organization's headquarters in Moscow. After finding out the information he needed, Atlis thanked the Memorial volunteer at the Moscow end of the telephone line and asked what her name was. She told him: Susanna Solomonovna Pechuro.

"*Susannochka!*" he gasped. "You're alive!"

Two weeks later, Atlis made the 4,000-mile flight across the country to Moscow, and the two of them stayed up most of the night talking. They had never met before. But four decades earlier, at the beginning of the 1950s, their lives had been fatefully linked. Susanna Pechuro was a member of a group of six people, arrested and charged with plotting against Stalin. Atlis had merely been friends with another person in the group, but because of this, he had gone to prison himself.

In the long years when Stalin murdered, by the millions, members of alleged conspiracies against him, almost all of them imaginary, Pechuro's group was a real conspiracy. And it was in the heart of the empire, in Moscow. Most extraordinary of all, its six members were all teenagers. At the time of their arrest, the youngest was sixteen, the oldest nineteen.

Three were shot.

Two died in labor camps or afterward.

Susanna Pechuro is the only one left alive.

One morning I knock on the door of an apartment more than an hour away from the center of Moscow by bus and subway. The woman who opens it is stocky, robust, about sixty years old. She walks with a slight limp. Her face is the picture of warmth, vitality, and luminous intelligence.

Pechuro's children are grown, she explains, but she likes to be around young people. So she rents out rooms in her apartment to students and young friends. In the living room Pechuro introduces one of her tenants, a woman named Lena. Her arm is in a sling— she hurt it yesterday in a karate class. She is lying on a couch, reading, covered with a blanket; Pechuro has obviously been taking care of her.

Before we sit down, I ask Pechuro about some framed photographs on a bookshelf.

She picks one up. "Here's a picture of the boys, my friends, who were with me in this organization. And who were executed. These three boys." We take seats at a table, and I ask her to start at the beginning.

"Most of us were high school or college students—first year at [Moscow State] University or at one of the institutes. What had brought us together? When we'd been younger, we'd all belonged to the literary club at the House of Pioneers. There we'd become friends. There we'd begun to think for the first time, and to discuss with each other what we were thinking about. We trusted each other completely. And in that we were right.

"We were forbidden to read our poems out loud unless they were checked by the director. We got indignant. There was a girl in our group who read aloud a mournful poem, about a girl who feels sad at a party because her boyfriend is dancing with another girl. She was told that this kind of mood wasn't worthy of a Soviet youth. She wasn't allowed to read this poem aloud. This was the last straw, and we announced that we weren't going to go to the club any longer, we were going to study by ourselves.

"In a civilized society, what we accomplished would be regarded as a trifle. What did we manage to do? Practically nothing. We issued two leaflets. We developed a program. The program said that in Russia we did not have a dictatorship of the proletariat—instead, we had a Bonapartist regime headed by a dictator. The program said that there were two imperialist systems that had divided the world into spheres of influence. That there was serfdom—although it was called collective farming. That all the officially proclaimed principles were being carried out in reverse. And that all this should be fought against."

Although their circumstances were incomparably more oppressive, the psychology of this little group sounds similar to that of the generation of the early 1960s in the United States, measuring the ideals of the Constitution and Bill of Rights, as taught in school, against the reality that blacks in the South could not vote. "We took on trust what they'd taught us," Pechuro continues. "When they talked to us about heroism, honor, and selflessness, etc., adults didn't mean anything by it. But we believed it. If you see that the world around you isn't fair, by all the rules you were taught at school, then you are honor-bound to fight to change it. This was straightforward, childlike logic. At my interrogation when they asked me who had taught us everything, the answer was, '*You* have. . . .' "

Did she and her companions know the risks they were facing? Yes, says Pechuro, "we knew what kind of country we lived in. We understood that any day we could be arrested. We knew that people had been arrested for less. If we were ten, by the time we were arrested, we would have let the truth be known to twenty. And they would spread it to another hundred. Our task was to get this process going, to start explaining the truth, not to be intimidated. Each of us realized very well what was in store for us."

The informal leader of the little group was a boy named Boris Slutsky, whose father had been killed in the war. "From the books left by his father, we figured out that probably he was a Trotskyist, who totally by chance had escaped being arrested. In one volume by Lenin, we found Lenin's will." The will warned Party members against Stalin. Although it had been distributed to some officials after Lenin died in 1924, Stalin had done his best to seize and destroy all copies of the will as he gained power over the next few years. The bookshelves of Slutsky's dead father also contained poetry by various banned writers, and this work, too, influenced the group of teenagers.

"This boy lived alone, because his mother had remarried and he refused to live with his stepfather. He had a room that no adult visited. There we gathered, there we talked, there we typed. We studied Marxism. Boris Slutsky was well read, bright and educated, in a way kids at his age usually are not. He explained that unless we had knowledge, we had no right to make judgments. We had assignments to read an article or a book, make a synopsis, and then we came together and talked, bringing our synopses.

"We read memoirs of revolutionaries. We made a hectograph—a primitive printing machine. We found a description of this machine in the memoirs of Vera Figner [a nineteenth-century revolutionary]. It could print 250 copies of something. We had all read books about members of underground organizations. They were our heroes. So we imitated them."

A teakettle whistles in the kitchen, and Susanna Pechuro gets up to fetch the tea. Lena, the young woman on the couch, has put down her book and is listening intently. Pechuro returns with the tea, and resumes talking. Her group learned many things from nineteenth-century revolutionaries, she says; one was caution. "I'd go to school and talk with my schoolmates. They'd think over what I said and agree. But we didn't invite just anybody into our organization, because we realized that this invitation could mean death.

Days would pass, and a boy or girl would say, 'Listen, if what you say is so, then we have to do something.' My answer usually was, 'No.' The boy would try to persuade me that something should be done. I'd say, 'Do you understand what the consequences could be?' It would go on for a long time."

Finally, Pechuro would say, "If you think this way, I can tell you that such a group already exists. If you want to, you can join."

"None of the people with whom we talked told on us. None. When they speak about mass betrayals, mass cowardice of the entire nation, I can say that it was not so among our friends."

Pechuro's group formed in 1950 and met at Boris Slutsky's room every day or two, after school. The authorities were watching them, for they had been overheard talking about forbidden poets. Slutsky himself had been under suspicion ever since, *as an eighth-grader,* he had made a speech at a Komsomol meeting protesting the famine that swept the western regions of the USSR in 1947.

"They started to follow us," says Pechuro. "They didn't bother to hide it. I'd be going to school and two men would be following me. We weren't afraid at first because we were too young. But when this circle around us began to tighten, then we got scared. Saying goodbye each time, we realized that we might not see each other again. We knew about the tortures. But it was one thing to know what was in store for us, another to feel it."

In 1951, everyone was arrested. The core group was only six, but eventually the police seized seventeen young people in all, "the others because they were our friends." One of the others was Miron Markovich Atlis, the man who had been so overjoyed, nearly forty years later, to learn that Susanna Pechuro was still alive.

Another of Pechuro's tenants comes out of a back room in the apartment, drawn by the sound of Pechuro's voice, and joins us at the tea table. She is a young woman of university student age with Central Asian or Tartar features. She, too, apparently has not heard the full story before, and listens quietly, sometimes asking a question.

At the time of her arrest, Pechuro was living with her parents and brother in one small room of a communal apartment that held seven families. Her parents had no idea what she had been doing. "When they came to arrest me, my parents were shocked. When they were taking me away, they let me say goodbye to my mother. I whispered to her to clean everything up because during the search I had managed to hide something. I needed her to remove this. She flinched. At that moment she understood."

Before her arrest, had she ever discussed her political ideas with her parents?

"Never. Because they would have gotten scared immediately. My mother has now died; my father is still living, and he says, 'It's frightening for me to die cheated. To think that I've spent my life being cheated. That I still believed—at a time when *you* understood everything. And I, a grownup, believed. My life has been spent in lies.' That's what he told me in the hospital recently."

For her parents' generation, Pechuro says, "The war [World War II] provided explanations for everything. Things were bad because of the war, what could you do? They hadn't traveled anywhere. They had read very little. They were people who went through their daily routine, brought up their kids, cooked their meals, went to their offices and didn't give anything else a second thought. If a newspaper said something, that meant it was so.

"Now people say they didn't see, then. But I think they really saw everything. It was easier for them to *persuade* themselves that they weren't seeing—then they were relieved of any obligation to do anything."

Pechuro gestures at her overflowing bookshelf. "Here I have books, reminiscences of women prisoners. Each starts out something like this: 'I was happy, I didn't understand anything. Then all of a sudden they [arrested me] ...' But excuse me, why was she happy? Why didn't she understand anything? Collectivization had happened. Millions of people had died of starvation. How could you be happy? There is a kind of poeticizing of this non-understanding."

The doorbell rings: it is a young man who has come to visit the ailing Lena. He wears a lapel pin with the Russian Republic flag—an informal badge, in these early months of 1991, of opposition to the Communist Party. He sits next to Lena on the couch and talks quietly with her for a few minutes. Then, overhearing Pechuro's story, he, too, joins us at the tea table.

By the time Pechuro was seized, in 1951, the NKVD had some two decades of experience in uncovering non-existent conspiracies. But they were remarkably inept at basic police work. An entire generation of investigators had been trained not in getting the facts, but in making people confess to imaginary ones.

"Under their noses, during the search I managed to hide a copy of our program. I had two copies." While crying and protesting that the officers had ruined the paper covers of her schoolbooks in searching them, Pechuro managed to move one copy of the program from the pile of material they had not searched to the one they had.

All Pechuro's group had done was to distribute a few leaflets. But this was so rare an act of defiance that this tiny collection of rebel teenagers, inspired by Lenin and forbidden poetry, was shunted up the line to the very top of the secret police apparatus. At one point Pechuro was personally interrogated by Victor Abakumov, the Soviet Union's minister of state security.

Pechuro is reluctant to talk about the means of coercion the interrogators used. "I don't think it's important how a person was beaten. No details are needed. To hit someone *just once* should be beyond the limits of civilized human conduct.

"I was very scared. You shouldn't think that we were made of iron. We were just kids. In my cell I wept a lot."

It took her some time fully to understand that she was not in a place like the Tsarist jails she had read about in revolutionaries' memoirs. "I was put into a cell with six other women. The first thing I asked them was whether you could escape. They looked at me, surprised. That was our level of knowledge, you see: on the one hand, we knew what could happen to us; on the other, it was all some kind of childish game.

"The women taught me a lot, however. They taught me how to make a needle out of a straw from the broom or a fishbone, how to take thread from a handkerchief or dresses, how to mend a hole, how to pretend that you were awake when sleeping during the day. They taught me how to tap on the wall to the next cell, showing me all three ways of coding the alphabet. When I was transferred, I was an experienced prisoner. I knew what to do."

Midway through being questioned, Pechuro had one of those extraordinary strokes of luck that usually occur only in novels. She believes it saved her life, and the lives of others. A new interrogator was assigned to question her—and she knew him.

"He recognized me. Because he had lived in the next apartment building to ours. He was ten years older than me. He remembered me as a small girl playing in the yard. And that somehow changed him.

"His name was Kim Smelov. Thanks to him, some twenty additional people were *not* arrested."

Smelov questioned Pechuro in a peculiar way. When he asked her about people she had known and what they had said, he would spread out the papers on his desk in a way that allowed her to read them. They summarized other interrogations and showed the extent of police knowledge of the case. In this way, she could supply answers that did not incriminate anyone the police did not already know about.

What lay behind Smelov's unheard-of kindness? Pechuro got a clue during one of her last interrogation sessions with him.

"He said, 'Will you teach me how to tap [on the cell walls]?'

"I asked, 'What for?'

"He answered, 'I think I may need it.'

"I said, 'I'll teach you.'

"He was arrested very soon after that . . .'"

After a long investigation, the three young men judged to be the ringleaders of Pechuro's group were shot. Almost all the remainder of the seventeen arrested—both group members like Pechuro and those guilty only of being friends with one of them, like Miron Markovich Atlis—got long sentences and were scattered into the *gulag,* some as far as Kolyma.

Pechuro pours more tea and brings out cookies. It is now almost noon, and we have been talking for two hours. The three young people are still listening, rapt, to every word.

Pechuro spent some five years in prisons and labor camps; she was among the last Stalin-period political prisoners to be released. "I was lucky. Because they were always taking me from one place to another, I saw a lot, I met a lot of people. I had a huge experience, and I think God somehow intended to make me remember it all, to tell it all later."

Much of her time in camp she spent just south of the Arctic Circle. Occasionally she would run into others arrested as part of her case. "Sometimes we would be together in the same camp. We did our best to try to help each other. Those who are alive now, we still meet. But not everybody, because attitudes vary. One woman says, 'I ask you not to remind me about all this. If I keep remembering it, I can't continue living . . .' Can you blame her for that? No, you can't. But those of us who remember our younger days with real happiness, we get together all the time. Two days ago at my friend Irina Orginskaya's, we sat laughing, remembering things from camp, and her daughter said: 'My God! One would listen to you talking and think you should be grateful to our government. You've got all that to remember, and we don't!'"

After her release from labor camp in 1956, Pechuro studied Russian medieval history, the time of Ivan the Terrible—the Tsar whose secret police was in a way the model for the NKVD. But the only work she could find was as a librarian. "All my life I wanted to be a teacher, but they wouldn't let me do it because of my prison record."

When the dissident movement of the sixties began, her children were small, and so, she says apologetically, she only "read *samizdat,* typed what I could, distributed literature, raised money, kept kids at my place when their parents were arrested. And so on."

Sitting at the tea table and hearing Susanna Pechuro talk, I am astonished that her story is so unknown. The Great Silence silenced tales of bravery as well as of suffering. But now, when all can finally be written and said, why isn't this woman a national heroine? Why aren't there books and movies and TV shows about her? Why, in what they read or watch about Stalin's rule, are most Russians interested in hearing only stories of tragedy? Perhaps the courage of Pechuro's little band of teenagers is a reproach to those who did not speak up. Pechuro's group bears a striking resemblance to the White Rose—the small circle of anti-Nazi resisters in Munich in 1942–43. They, too, were students who published leaflets, and most paid for their bravery with their lives. Their existence, too, was a reproach to their countrymen and women who stayed silent. It took forty years after the war before a German filmmaker made a movie about the group.

The lack of public recognition is the last thing on Pechuro's mind, however; she never mentions it. One invigorating thing about hearing her, and something that sets her aside from most Russians of her generation, is that she has always thought of herself as a political activist, not a helpless victim. She seemed to carry no emotional scars from her time behind barbed wire; instead, it was something she felt proud of. Most other camp survivors I met, if they drew any meaning from their ordeal, came away from it with illusions lost or lessons about human nature learned, like survivors of a hurricane who now know what their houses and their spirits can and cannot endure. By contrast, today Pechuro seems above all a happy person, radiating sturdiness, energy, and conviction.

Pechuro's activist stance gives her a refreshingly different eye on Soviet history. She is less interested, for example, in documenting deaths and suffering than she is in the rebellions that shook the labor camps a year after Stalin's death. The dictator was gone, but the prisoners had still not been released. Although never mentioned in the press, word of these rebellions spread through the country, and even more rapidly through the camps themselves. One former prisoner I met learned the news from urgent, desperate messages chalked on the inside walls of freight cars by inmates in other camps who had loaded or unloaded them.

"In 1954, rebellions involved thousands of people in some of the biggest camps," Pechuro says. "I was in a camp that didn't have a rebellion. I feel sorry, because one should go through this! One should see it with one's own eyes. In Kengir they stopped the rebellion with tanks. They drove the women between the barracks and let the tanks ride over them. In Norilsk, they bombed the camps. In Vorkuta, they simply shot people, en masse." In Kengir, one of Pechuro's high school friends who had been arrested with her joined the rebellion: for forty days inmates took over a big camp complex and set up their own newspaper and theater. When Red Army tanks came in, seven hundred prisoners were killed.

Few of the reformist historians I talked to were interested in the camp rebellions, and several dismissed them as tragic but politically ineffective. But the changes that came after Stalin's death, Pechuro asserts, were not merely reforms handed down from on high; they were something people fought for in these uprisings. "All the 1956 reforms and the shutting down of the camps were caused by those rebellions! It was no longer possible to keep this army of people in obedience. When the camps rebelled, coal-mining output dropped, timber-cutting also. Nobody was at work. Gold and uranium—no one was working. Something had to be done. Nikita Khrushchev released us. What else could he do? *We managed to make them release us.*

"It's the same thing now when they say that *perestroika* is Gorbachev's child. No, it's not Gorbachev's child! Having come to power, Gorbachev had wits enough to understand that things could not go on the way they had been. But this has been the victory of the dissidents. Each dissident had two hundred sympathizers. They taught people how to think. They defied fear. Those eight people who went to Red Square [in 1968, to protest the Soviet invasion of Czechoslovakia] made millions stop being afraid."

Some time ago, Pechuro left her librarian's job and went to work as a night-shift *dezhurnaya,* or floor matron, at Moscow State University, so as to have her days free to take care of a grandchild with a chronic illness. But even after she had earned her retirement pension, she stayed on at the University. "Because I like it there. The students know me, it's a pleasure to spend time with them. They come and talk with me during the night and ask my advice. Maybe this is the never-happened teacher inside me . . ."

All her spare time, Pechuro spends working for Memorial. She has drawn many young volunteers into the organization, who help the old and the sick. "They visit blind [survivors] and read aloud to them. They go to meetings and demonstrations. They sort files and fill in index cards. Their age range is from eleven to eighteen. Now we've set up a Committee to Abolish Capital Punishment. And these kids were the first to join!"

On Pechuro's crowded bookshelf I notice a photograph of a teenage girl, her hair in a short braid, wearing a dress with a narrow, prim collar with rounded corners. Her face has a look of determination and intelligence, and a certain kind of beauty that flows from that. Was this another member of the group? No, Susanna Pechuro says, it is herself at age seventeen, some three months before her arrest.

I am so struck by this photo that I take it off the shelf and leave it propped against the teapot, a new, silent participant in our conversation. As the talk continues, I glance from the photograph to the warm, animated face of Pechuro speaking, to the alert faces of the three young people, still listening intently. For a moment I feel something almost transcendent. It is as if the barriers between the living and the dead have dissolved, and a spark of something—the example of integrity, the power of truth, proof that in the very worst of times it is possible to see everything and to deny nothing—is being passed from this determined young woman in the photograph and her dead companions to the people in this room. It feels an honor to be a witness.

Secret police orphanage for children of dispossessed *kulaks* and other "enemies of the people," western Siberia, 1930s. The official on horseback wears a revolver.

4

"Why Are
We Weeping?"

After some weeks in Moscow, I began to feel impatient to see a piece of the old *gulag* itself, that territory of land and spirit that was so much a presence in the minds of people I had talked to. *There,* was how survivors referred to that world: I came back from *there*—as if from a place off the face of the earth. From the work of photographers who arrived with the liberating armies, the sight the Nazi death camps is branded on our minds. But we have no such familiar image of the Soviet *gulag,* and I wanted to see what it looked like.

The old camps of Kolyma would still be buried under the polar winter's snowpack; seeing them would have to wait a few more months. One day, however, I noticed a newspaper item from the city of Karaganda. In the *gulag* years, the Karaganda area's huge camp system was close to that of Kolyma's in notoriety. It was spread over a broad expanse of the steppe, or prairie, where Siberia meets what used to be called Soviet Central Asia. The labor camp in Alexander Solzhenitsyn's *One Day in the Life of Ivan Denisovich* is modeled on one near Karaganda. The city's "tour bureau," a paragraph in the newspaper said, was now offering visits through the remains of the regional headquarters of the camp system.

Our son, who was going to the American school in Moscow, had his Easter vacation coming up, and so we planned a week's trip—to Karaganda and to Siberia's largest city, Novosibirsk, not far away. Karaganda is nowhere near the usual tourist track, and its tour bureau, whatever that might be, was not part of the big state agency for foreign travelers, Intourist. On the telephone, the Karaganda people sounded delighted that an American family wanted to take their tour of the old *gulag* buildings. Were we interested in other scenic attractions? No, I answered, but could they find me a former prison camp guard? They said they would work on it.

The steppe around Karaganda is a wide, dry expanse, which a century ago was peopled mainly by nomads herding their camels and sheep between water holes. In recent years, the Soviets have used these open spaces for launching space rockets and testing nuclear weapons. Geographically, Karaganda is on the southern edge of Siberia; politically, it is in Kazakhstan; ethnically, its inhabitants are mainly Russians.

Karaganda sits on top of one of Siberia's largest coal fields. It was to sink the mine shafts, dig the coal, and lay the railroads to get the coal out that prisoners were brought in the early 1930s to what had previously been only a small village. These laborers and their masters had to be housed and fed, and other prisoners were put to work quarrying, building, and working the region's new collective farms. By the time of Stalin's death, about a million convicts had passed through the Karaganda area *lager,* or camp, system. This was known, with the Soviet taste for acronyms, as Karlag.

Prisoners released from camp usually had to remain some years more in exile. For former Karlag prisoners, this often meant working in the factories of Karaganda. It was a drab, hastily built city, heated, powered, and polluted by the local coal, and plagued by shortages. "Distances in Karaganda are great," writes Solzhenitsyn of his own exile there in 1955,

and many people had a long way to travel to work. It took the tram a whole hour to grind its way from the center to the industrial outskirts. In the tram opposite me sat a worn-out young woman in a dirty skirt and broken sandals. She was holding a child with a very dirty diaper, and she kept falling asleep. As her arms relaxed, the child would slide to the edge of her lap and almost fall. People shouted, "You'll drop him!" She managed to grab him in time, but a few minutes later she would be falling asleep again. She was on the night shift at the water tower, and had spent the day riding around town looking for shoes—and never finding them.

That was what exile in Karaganda was like.

Before leaving for Karaganda, I wanted to talk to survivors. I found several in Moscow. Saul Shuv was one. A retired mechanical engineer of eighty-one, he had been arrested in 1950, and was sent to Karaganda after a five-minute trial. He spent the next five and a half years as a prisoner there, working in a meat-packing plant, a brick factory, and a sawmill. He vividly remembers going to work in columns of several hundred convicts, surrounded by guards, "with dogs and armed, of course: 'Step to the left or step to the right, and we'll shoot!' "

Shuv spoke slowly, with the anxious methodicalness of an elderly person afraid that he might forget something. He insisted on running through his life story from the beginning, naming all his brothers and sisters, mentioning every date and detail in his arrest and interrogation. It was as if he meant this accretion of fact to prove beyond any doubt that he was a loyal citizen from a loyal family, completely innocent of the charges against him.

I heard these protestations of loyalty and innocence from many *gulag* survivors. They were understandable and no doubt true, but after a time they always began to feel to me like something of a denial—that the entire vast machinery of state terror was irrational, that loyalty didn't count, and that in a sense everyone was innocent because there was no organized opposition to the state. And that exactly that absence was the problem.

Shuv showed no bitterness. He talked about his time in the *gulag* not as if it were a monstrous wrong done to him, but as if it were inevitable, like an epidemic or a drought. In him, the Great

Silence about Stalinism had repressed not the memory of what had happened, but any feeling of the right to be angry about it.

He spoke proudly of having fixed a key piece of equipment at the sawmill in Karaganda, and of how he had worked at the meatpacking plant "on the heating system, the water supply system, the sewage and the ventilation. I was an ordinary worker, but the team leader understood I was an engineer, and when there was an important task, I was in charge of it. I worked hard." He seemed eager for the approval of the team leader then, and of me now.

Shuv's lack of anger at first seems puzzling, but it is not so rare. Stalin had all his open opponents shot in the 1930s; the millions of Soviets in prison camps at his death in 1953 had mostly grown up, like Shuv, as believers. "We were brought up, you know, in the spirit that Stalin was God," said Shuv's wife, Lena. "When we learned the truth, it was hard on us."

After Stalin died, conditions in the *gulag* got a little better. Shuv's wife was allowed to come out from Moscow to visit him briefly. Thousands of *kulaks,* the peasant farmers forced off their land by the collectivization of agriculture, had also been deported to Karaganda. "When I visited," she said, "I went to buy some milk outside the camp. Former *kulaks* were living there. When I said that it was a convict I needed milk for, this woman poured me almost all cream. . . ."

Another former Karaganda prisoner, Olga Infland, also remembers the kindness of the *kulaks*. Her ten-year prison term expired in 1947, but she had to live on in Karaganda as an exile. She and a friend, both released from camp, searched desperately for housing. "Dispossessed *kulaks* had built their homes there. We came to one house, to another, asking whether there was a room for rent. A guy came out and said, 'Are you from the camp?'— which meant we were okay. 'My son has a room,' he said, 'I'll ask him to let you have it.' And so we went to live there. Alexei Nikolaevich Zlobenko—may his memory live forever!"

Olga Infland was the most articulate of the former Karaganda prisoners I talked with, although, like Shuv, she told stories of suffering and of endurance, not rage. An eighty-three-year-old retired literature teacher, she had broken a rib recently, and winced when she moved, but she smiled easily and overflowed with Russian hospitality. She knew that foreigners often get lost among Moscow's confusingly numbered apartment buildings: "Next time tell me when you're coming and I'll meet you down on the street and guide you in."

When the firestorm of the Great Purge swept over the Soviet Union, Infland was at risk because she had been born abroad. Until she was in her twenties, she lived in the large Russian community in Harbin, China. Many Harbin Russians were railway workers; Harbin was the major city on an arm of the old Trans-Siberian Railroad that takes a short cut to the Pacific by slicing across Manchuria. In the 1930s, the Japanese took over Manchuria and the Soviets sold them the railway. Infland and her family, along with thousands of other Russian *Harbintsi,* left for the homeland many of them had never seen. A few years later, in 1937, she was arrested.

Interrogators tried to force Infland to sign a statement incriminating the rest of her family as spies. She refused. She was put through a mock execution: a soldier with a revolver marched her down to the prison basement and told her to stand against the wall. "Probably you are not interested in all that," Infland said, as if she didn't feel the right to dwell on it.

Like Saul Shuv, Infland accepted her fate with a curious resignation. Her tone was not one of bitterness, nor of a desire for justice, nor of burning curiosity about what made her country go mad. Unlike various other *gulag* veterans I had met, I found Shuv and Infland through mutual friends who knew I was heading for Karaganda, not through an activist group like Memorial. I suspect their quiet sorrow is typical of many Soviet prison survivors, especially those in old age, who don't want to join anything, write anything, or rock any boats.

Years after finishing camp and exile, Infland tried to figure out how many people she knew personally had been shot or died in prison camps: "I made a list of some forty-nine people." Ironically, the Harbin Russian community had been politically split: thousands of the Tsarist loyalists eventually wound up in San Francisco (where, decades later, one gave me Russian lessons); those like Infland's family who returned to Russia were generally the Communist sympathizers.

As a schoolgirl in Harbin, Olga Infland had even belonged to the Komsomol, or Young Communists. "Some fifty years after my arrest, two or three years ago, I was invited to the Komsomol Central Committee and they gave me that"—she pointed wryly to a certificate lying on a bookshelf—"and a badge for my Komsomol activities. Fifty years after. They gave awards to four people. Four people from Harbin who survived."

Infland showed me her collection of old family photos, which miraculously had also survived all these years: children on a lawn in the vanished world of Harbin; young men posing proudly in photographers' studios; demure young women in formal dresses; two young male cousins laughing and clowning at a dining table; Infland as a radiantly beautiful young woman, photographed by her husband. Leafing through the photos, she pointed at each person: "Shot . . . shot . . . shot . . . shot."

Infland was one of the first in her family to come from Harbin to Moscow; today, nearly sixty years later, it still weighs on her that some of the others came because of her. "I invited my two brothers, I persuaded them, I talked them into coming."

Infland's husband and one of her brothers were arrested before she was, and the *Harbintsi* in Moscow sensed a cataclysm closing in on them. "I headed to an attorney. I was not the first. There was a big room, filled with people.

"He asked, 'What's happened?'

"I told him, 'My husband has just been arrested.'

"He said, 'They've already started to arrest wives. If you want my advice, take your mother and get out of town.'

"I persuaded my mother, and she bought her ticket for September 11. On September 10, we were arrested. She hadn't wanted to leave earlier because my brother's birthday was approaching and she felt certain that he would be released on that day. 'He'll come and I won't be there!' she said. That's why she stayed.

"All of us were arrested, nine people. My husband and I, my mother, two brothers, another cousin, my mother's cousin and his wife, and another cousin, Sasha. Two people returned—Sasha and I. All the rest were executed, because we had lived in Harbin. All of them. Sasha and I met each other in camp, in Karaganda."

As we flew to Karaganda, below us the snowy Siberian evergreen forests gave way to endless, bare steppe. By mistake we got off the plane for a moment at an intermediate stop, a spooky, desolate spot with a strong wind, a row of crop-dusting biplanes, and no trees or town in sight.

At the Karaganda airport, a short, prim woman from the tour agency was waiting for us. She introduced herself in Russian fashion, by first name and patronymic, Marina Nikolaevna. As I stood talking with her for a moment, another woman—larger, heavily made up, and breathlessly eager—came rushing up. Hearing me, she burst out, in English, in a tone of transparent disappointment, "Oh! You speak Russian!"

Some of the patient people who have tried to teach me the language over the years might consider this an overstatement. But it was true enough so that I had been doing all my interviews in Russian, and had not asked for an interpreter in Karaganda. That was the job that this second woman, who introduced herself as Ludmilla Nikolaevna, had been hired by the tour agency to perform.

We didn't have the heart to turn her away. Driving into town with the two women, we could see that Ludmilla Nikolaevna was in a state of great excitement. Voluble, extravagantly friendly, and wearing spike heels, she was a divorcée in her mid-forties who taught English at a local military academy. This was a great day in her life, she said: we were the first people from an English-speaking country she had ever met.

Our visit to the former Karlag headquarters was set for the next day, Marina Nikolaevna and Ludmilla Nikolaevna told me; they would come for us in the morning. Now they dropped us off at our hotel, an enormous, gloomy building where the elevators were not working and there seemed to be few guests. The desk clerk gave us food coupons for the dining room: outside Moscow, food is often rationed. In the empty lobby a huge electric shoe-shining machine stood high as a person's head; it swallowed our coins but would not turn on. On the hotel's main staircase lay a dead fish.

During the evening, we explored Karaganda. The most notable building in town was the Palace of Miners—a ghastly columned structure with pseudo-Islamic concrete grillework, topped with huge statues of a miner, a collective farmer, a soldier, a woman worker, and other heroic figures. A nearby building had an observation deck on the roof. The skyline we could see from it was filled with black piles of slag from the coal mines, and a distant forest of smokestacks from a big iron and steel works that spreads across several square miles on the outskirts of town. In the graffiti scratched on the observation deck's parapet was a dollar sign.

Karaganda was built largely by convict labor. Reminders of this keep turning up: when a construction crew was renovating a building here recently, they found some letters hidden behind bricks in a wall, which prisoners had left nearly forty years earlier, hoping that a "free worker" would find and mail them.

The city has a population of some 800,000 today. Unlike Moscow, where they have been taken down, the old slogans were still on billboards in giant letters on apartment house roofs: GLORY TO THE MINERS' LABOR! GLORY TO THE SOVIET PEOPLE! GLORY TO THE FIVE-YEAR PLAN! Lenin, in cement or bronze, was everywhere: resting his arm on a lectern in a hotel lobby; rising, in a cape, out of a block of pink marble on a public square; and gesturing from a pedestal in front of City Hall. At the railway station, peasant women with bandanas and bundles crowded the waiting room. The constant wind blowing from the steppe filled our hair and ears with a fine dust.

Although Karaganda was built in the 1930s, many of its gray, standard-issue Soviet apartment blocks are much newer. There is a reason for that. When you remove large amounts of coal from the earth, often the surface of the ground eventually subsides. No one, however, dared make allowances for this when hundreds of thousands of prisoners were digging coal back then, desperately trying to meet the rising quotas of the Five-Year Plans. Mine managers who didn't meet their quotas risked jail, or worse. At all costs, the mines produced the coal.

The bill came due some several decades later. By the 1960s, much of Karaganda was slowly sinking. Water and gas mains broke, and many apartment houses, streets, and a major bridge began to collapse. Trolleys on the way to the railway station had to stop short for fear of sliding below ground. More than 300,000 people had to be moved into new housing. The Soviet Union is not the only country where mining has led to environmental catastrophe, as the scarred hills of West Virginia testify. But here, as all over the USSR, deathly fear of not meeting quotas had meant that this supposedly planned economy had really not been planned at all.

The next morning, the two women picked us up at the hotel in a small van. As a sign of our VIP status, it had curtained windows. The cast of characters for our long-awaited tour to the old labor camp headquarters had overnight ballooned to a total of nine people. On board the van were myself; my wife, Arlie; our fourteen-year-old son, Gabriel; Ludmilla Nikolaevna, the English teacher; Ludmilla Nikolaevna's teenage son; Marina Nikolaevna, from the tour agency; Marina Nikolaevna's seven-year-old daughter, dressed in her Sunday best, with a large white bow in her hair; the van driver; and finally a tall, intent frowning man introduced to us as Nikolai Fyodorovich.

The unexpected Nikolai Fyodorovich wore a black homburg and a long black trench coat. He seldom removed either, although the day was warm. He was, Marina Nikolaevna explained, *spetsialist po lageram,* a "specialist in camps," a retired history teacher.

We were the first foreigners to take this excursion, and to guide us through it, Nikolai Fyodorovich held on his lap a thick notebook. On the left-hand pages were pasted newspaper clippings; the right-hand pages were his own handwritten notes, which he recited, for most of the day, in booming, sepulchral tones. Often in pre-*glasnost* days I heard official Soviet guides mechanically recite their city's heroic achievements in producing machine tools or truck engines. Nikolai Fyodorovich had the same solemn cadence, but his statistics were all about prisoners, executions, and exiles, the number of Karlag camps (twenty-six), and the size of the area covered by its slave labor economy (as big as France). Every half hour or so, all day, he would lean forward from the seat behind me and anxiously tap me on the shoulder. He would point to the clippings on the left-hand pages and say, "It's all documented! All documented!"

When he was not talking, Ludmilla Nikolaevna eagerly took over. She was dressed for a party, with rouge, lipstick, dyed black hair, and small bits of jewelry everywhere. She had fitted her ample shape into a tidy gray suit that looked businesslike enough but revealed her cleavage through a sheer blouse when the jacket pulled open. Thrilled that she could use her life's work, English, on native speakers for the first time, she boldly seized the chance to try out all these long-stored words. She launched enthusiastically into the story of how they discovered coal here: "A hundred years ago, a cowboy discovered a lump of black! He put it in the fire and it burned! Then the cowboy ran to tell his master. . . ."

We passed through the sooty factory belt on Karaganda's outskirts, and soon were on the steppe beyond. The coal discovered by that cowboy, whoever he was, lay everywhere in the ground beneath us. In the old days, coal from here had been unusually cheap because, of course, these miners didn't have to be paid. Karaganda coal fueled much of the Soviet Union's crash industrialization of the 1930s and after. Now and then we passed a strange reminder of that era, a place where the ground had collapsed above an abandoned mine: a field, and sometimes a shack or fence as well, would be covered with water from the end-of-winter snow melt.

What was it like working in those mines? "Sometime in late

October or November 1953," writes one former Karlag prisoner, Alexander Dolgun,

> there was a series of power failures while we were below ground. The elevator was electrically powered. This meant a climb by ladder, 240 meters down in the morning and 240 meters up at night with no light but the meager carbide lamps in our helmets. The rungs of the ladder caught the drip from the walls of the shaft, and where the cold air swept down from outside it froze quickly so that the climb was a terrifying, exhausting nightmare in which you might lose your grip or your footing on the icy ladder at any moment, and the bottom of the shaft was a thousand feet below ground. There were several screaming, gyrating falls to the bottom, and even the guards found the whole experience unnerving.

The arid brown plain we drove across had the feel of the endless cattle rangeland of the American West, where the horizon is far off and seems to recede before you forever. The lack of trees makes for high winds and a disorienting absence of landmarks. In winter, there are severe blizzards.

"Many people got killed in those storms," Olga Infland had told me in Moscow. "When the storms came, they lasted for a day, three days, nine days. The barracks had to be dug out. We had a case when a storm came suddenly and two people were carried away."

Once, traveling in a horsecart, Infland got caught in a bad storm. "It was about six in the morning. There was only the steppe, and the steppe, and the steppe all around. No road at all. Suddenly I saw things shining. It was dark and I couldn't see what they were, but they were shining. When day broke, it turned out that wolves were running along close to us. These were their eyes."

As the van drove on, Marina Nikolaevna of the tour agency sat stiffly with her daughter and said nothing. Nikolai Fyodorovich intoned his statistics and the names of prisoners: "Amosov, Stepan

Stepanovich, writer, repressed here! Muralov, Vitaly Ivanovich, People's Artist of the USSR, repressed here!" Ludmilla Nikolaevna sat with Arlie and Gabriel and translated for them.

After a while I began to notice something. She wasn't exactly translating what Nikolai Fyodorovich said. Her account of things was always more colorful. When we passed the pitheads of several coal mines, where elevator machinery was housed in ramshackle wooden towers, I asked Ludmilla Nikolaevna in English if the coal miners around here were on strike, as they were that month in some other parts of the Soviet Union.

"Yes!" she said excitedly. "They've put down their shovels, put down their picks, and won't do any more work until communism has ended!"

The next coal mine we passed was close enough so we could see the large metal wheel of the elevator cable that hauled coal to the surface. It was turning. While Ludmilla Nikolaevna was talking to Arlie, I asked Nikolai Fyodorovich in Russian if the miners were on strike.

"No, they settled three weeks ago," he said. "The president of Kazakhstan made a deal with the union, and they're all back at work."

The same divergence appeared throughout the day. After an hour's drive, we reached the town of Dolinka, site of the old Karlag headquarters area. Our first stop was at the ruins of the system's central "isolator," the prison for interrogation and solitary confinement. The walls were still partly standing, but the roof was gone. In English, we asked Ludmilla Nikolaevna why.

"The people who worked here," she said with great passion, "the guards, the torturers—last year they came and destroyed the place so there would be no evidence of their crimes!"

While she and Arlie walked off to another section of the ruins, I asked Nikolai Fyodorovich the same question in Russian.

"The plaster here is mud-based," he said. "The roof started falling in after a bad storm recently, and the rain did the rest."

The prison building had been a large one; still visible in the remains of one yellow wall was a bas-relief of hammer, sickle, and star. We stepped over the remnants of the thick walls between cells, now mostly crumbled to knee height. In one tiny cell space we found a mat of woven twigs of *karagandik*—a wiry, thorny bush that grows on the steppe. "Some of the women slept on the floor on twigs," writes Margarete Buber, a German Communist arrested in the 1930s, of some time she spent in a Karlag punishment cell.

Dolinka is a rambling settlement of several thousand people, many of them workers from collective farms on the surrounding steppe. Along the town's roads, men and women trudged to and from the farms, wearing bandanas, high rubber boots, and felt jackets. Nikolai Fyodorovich pointed out long prisoners' barracks that have now been converted into workers' dormitories, and a white, columned building that was once the prison hospital. There were separate wards for guards and prisoners, of course, but both were taken care of by prisoner-doctors, one of whom had worked at the Kremlin hospital.

It was in this hospital in Dolinka that Olga Infland worked as a nurse for much of her time in Karlag. She said that once in this low white building she had been driven to despair by a harsh supervisor, "a voracious beast, a terrible woman." Infland tried to commit suicide by taking an overdose of morphine. She was caught in time, and punished—for unauthorized use of hospital property.

A different official, this one kind-hearted, saved Infland on another occasion here. "She was a very good woman, she protected us. She was physician-in-chief of the hospital. An order came saying that all Article 58 people [political prisoners] should be transferred from Dolinka to a remote district three hundred kilometers away, with bare rocks and no trees, a most terrible place. She summoned me and said, 'Take these four patients now, occupy two wards, and lock the doors. Don't let anybody in. If someone comes, tell them you have the patients with contagious diseases.' "

Continuing our drive around Dolinka, we passed one camp from the old days that, like a number of old *gulag* installations, is still in use as a prison today. Now it holds only common criminals, but hundreds of political, religious, and nationalist dissidents were confined in such places until the late 1980s. It was ringed by a long wall topped with barbed wire, with rusty metal guard towers at the corners.

Nearly a thousand babies were born to prisoners in Karlag, Nikolai Fyodorovich told us. Food was scarce, and many died. We stopped at a children's cemetery, where graves were marked with crosses of rusted iron. "N. I. Podoyesnaya, 18.4.49 to 19.3.50," read one. The metal crosses were mostly tilted or bent, sometimes flattened against the ground, by years of frost and thaw. Electric power lines crossed above the cemetery, a few early sunflowers dotted the sparse grass, scraps of plastic bags and refuse blew in the constant wind.

Close by were adult graves, reminders of how many countries were crushed between Hitler and Stalin. A cross stood atop the grave of a Polish officer; a plaque left by recent visitors marked a section of Lithuanian graves—some of the hundreds of thousands of people shipped off to prison or exile when the USSR took over the Baltic states as a result of the Nazi-Soviet Pact. The name Karaganda also carries a special echo in Spain. When Franco's Fascists won the Spanish Civil War, several thousand Spanish Republicans took refuge in Russia, which had backed their side for most of the war. They were paraded through the Moscow streets as heroes. Stalin, always suspicious of foreign leftists, then promptly dispatched almost all of them to the *gulag,* and most ended up in Karaganda. Many are buried here.

Dolinka was as flat as the surrounding prairie, and most of the homes were simple, metal-roofed shacks with privies and piles of coal in small backyards. But soon the van pulled up at what seemed like a majestic estate: a screen of trees, grass, and a stately two-storied building with two wings. A second-floor porch framed by white columns gave it the look of a grand Southern plantation manor. But a manor that had seen better days, for this mansion's garden was overgrown, its fountain dry, a pedestal bare.

Stalin's statue had once stood on the pedestal, and this building had been the administrative headquarters for all of Karlag. It was here that the records were kept—and burned, Nikolai Fyodorovich said, when Khrushchev shut Karlag down. It was from here that orders went out to the twenty-six labor camps of the system, and to here that their commandants came for meetings. After Karlag closed, the building was used for a children's sanitarium. Now it was being renovated yet again, but the work seemed stalled. Bricks, logs, and other debris were strewn in front, but we saw no workers. A woman's face peeked out of a side window, then disappeared. A guard? A squatter?

On a patio next to the building a prisoners' orchestra had once played at dances for the secret police officers and their wives. Gingerly trespassing, we walked up the wide front steps, along a high-ceilinged hallway, and up a flight of stairs to a second-floor room that looked out onto the columned porch. Everything was covered with crumbled plaster, but through the dust we could still see the handsome parquet floor. This office had once belonged to General Victor Zhuravliev, commander of all Karlag and a candidate member of the Communist Party Central Committee. Once the absolute ruler of this region, he, like so many top officials, later apparently became a victim. Several years after he occupied this office, Zhuravliev was expelled from the Central Committee and dropped from sight. There is no record of how he met his end. Was it in one of the very camps or prisons he had run? Did he think back, in his last days, to his parquet floor, his columned porch, his open-air dances on the patio?

Just beyond this headquarters building stood the remains of a school. Many *gulag* inmates were scholars or professors, and here some were put to work—teaching their jailors' children.

Most of the intellectuals in prison here never had such a chance. One Karlag veteran I met had been amazed when he was once transferred to a work crew whose other members were all academics: "Can you imagine? Thirty-six men, all Ph.D.'s, professors in various fields. We laid bricks."

Emerging from the grounds of the crumbling white manor house onto the dirt streets of Dolinka felt like stepping from the past back into the present. A bicycle race swept by us, the riders in helmets and skin-tight, brightly colored shirts: emissaries from the land of the living passing through the land of the dead.

What did the millions of prisoners locked up in places like Karlag feel at the time? Did they see their imprisonment as one of the great injustices of the century? Some did, of course, but what is more interesting is how many did not. And how many put a favorable gloss even on events they witnessed personally. Olga Infland, for example, told me about the executions announced to the prisoners: "At the beginning, every night the soldiers came and read the lists aloud: so-and-so, and so-and-so, according to such-and-such an article, executed." But she was under the impression—and still today seemed to be under the impression—that these were dangerous criminals who were being shot to make the camp safer "for us," the political prisoners. A large body of evidence, however, suggests that the "politicals" in the labor camps were far more likely to be punished or shot than the common criminals.

Like most inmates, Infland was fully aware of the *gulag*'s vast size. From an exile friend working in the personnel office, she knew how many camps there were in the region. And from comparing identification numbers with new arrivals, she could estimate how many thousand prisoners were coming into Karlag each year. But despite all this, during her imprisonment at Dolinka, Infland was one of those people who thought that "Stalin didn't know what was going on. That's why we were busy during those first years writing letters to Stalin. We were writing and writing and writing."

How could that faith be sustained through so many years in the camps—whose very existence, one would think, would undermine it? And how could anyone believe that Stalin "didn't know," when prominent Great Purge victims, thunderously demonized in every newspaper, included the great majority of the Red Army's top generals and more than two thirds of the Communist Party Central Committee?

In an intriguing little book called *When Prophecy Fails*, a team of American social scientists asks why, in religious sects that preach a Second Coming, the level of devoutness and proselytizing often increases when the Messiah does not appear at the predicted time. Their answer is that it is precisely the threat of a shattered belief system that gives the faithful a powerful new incentive to bond more tightly with each other, and to recruit new believers. Together, they can then convince each other of the necessary rationalizations (the Second Coming occurred, but in Heaven and not on earth; the delay is a part of a divine plan to test our faith; and so on).

There is an obvious analogy to the Soviet experience. It was, I think, precisely because people were suffering so much that they were willing to delude themselves and to deny the obvious. Without the denial, all the terrible suffering would have seemed purposeless. Many people wanted to believe that the camps, the deaths, the collectivization, the famine, the privation and shortages, like the suffering of a noble war, were for some reason necessary. That it was for the sake of progress, of industrialization, of a better future. That the nation was a family writ large and Stalin its stern but benevolent patriarch. And the worse the suffering, the more fervently the believers believed. For the greater was the potential abyss of despair if those terrible hardships were all for nothing.

Olga Infland was living in Karaganda as an exile, after ten years as a camp inmate, when Stalin died. Like many prisoners and exiles, she wept at his death. Today, she can scarcely believe it: "When Stalin died, there was a meeting, speeches were made, and we cried. Really, we were crying. Can you imagine? When we left the meeting, I said to my friend, 'Why are we weeping?'"

5

Lunch
with the
Colonel

Our drive around Dolinka took most of the morning. About
noon, the curtained van pulled up at a pleasant, spacious
home of six or seven rooms, behind a wooden fence, with a garden
and small orchard. "The only house for miles around with a flush
toilet!" said Ludmilla Nikolaevna. That may or may not have been
true, but it definitely was the biggest house we had seen in
Dolinka. Living in it was a former secret police colonel.

Someone in his position never would have been willing to talk
in more security-conscious Moscow. Given this rare opportunity,
Arlie and I agreed that she and the others would wait outside.
Whether proud or repentant, surely anyone retired from the concentration camp business would speak more openly if not confronted with our entire vanload of nine children and adults,
including the outspoken Ludmilla Nikolaevna.

So I left almost all of them in the van and went through a gate
in the fence, accompanied only by Nikolai Fyodorovich in his black
homburg and trench coat. To collect data for his notebook, he said,
he had some questions of his own he wanted to ask the colonel.

Colonel Mikhail Volkov was sixty-two, a large, strong man
with a square jaw, closely cropped gray hair, and an iron hand-
shake. He carried his extra weight as Russians often do, not as a
potbelly, but like a sweater, as more evenly distributed armor

against the cold winters. The colonel's shrewd eyes fitted his part, but I could also imagine them as the eyes of a Chicago policeman, or of a hard-bargaining farmer in some rocky, cold region of the United States or Canada. The most striking thing about him was his huge, leathery hands, the fingernails trimmed or bitten so far back they were only three quarters the normal length. What had these hands once done to prisoners? It was easier to imagine them pulling up turnips.

The colonel introduced his ruddy-faced son-in-law, a coal miner, and his wife, Olga, also retired from the prison system, a resolute-looking woman with hazel eyes. His daughter, the son-in-law, and their two small children all lived in the house with them, he explained. We sat down at a table in the living room–dining room, a large space with a red felt couch and a Persian-style rug hung for decoration, in Russian fashion, on the wall. Potted plants, a cactus, and some tomato seedlings sat on a windowsill.

I'm not sure what I had expected the colonel's house to be like, but I had certainly assumed it would look different from those of the democratically oriented writers and historians I had mostly been interviewing until now. It didn't. In several rooms there were bookcases; among their contents were the Russian classics, the works of Shakespeare in Russian, and a picture of the late Vladimir Vysotsky, the semi-underground poet-actor-singer who is the most beloved Russian cultural figure of the last several decades. (One of his most famous songs is about a true believer with a tattoo of Stalin on his chest, who gets sent off to Siberia nonetheless.) Only since *glasnost* have recordings of Vysotsky's gravel-voiced, harshly ironic ballads been fully available. Seeing him and Shakespeare together here was as unexpected as finding, in the home of a retired American prison superintendent, the complete works of Pushkin and a picture of Bob Dylan.

Colonel Volkov was posted to Karaganda in 1949, he said, as soon as he finished his training. He had served here all his working life. "I was sent to the political department the moment I came. I became deputy chief of the section, and then chief of the section. To the end of my tenure I was in political work."

What, exactly, did "political work" mean? The colonel's reply was almost impenetrable:

"Organization of public work, of working competition, competition for the best monthly results. We checked the results regularly. For teams and working departments, we did it monthly. Once a quarter we had a meeting of all working convicts. When detachments were introduced, we did it for detachments. Say, for example, to accomplish a harvesting campaign, or a sowing campaign. If it was harvesting time, we checked the results every five or ten days. People were always, so to say, in a mood to accomplish those activities. All the issues were discussed at the meetings of the political department. If the organization wavered in producing food, we improved the situation, discussed things, asked questions of the Party officials who had made work performance drop. When things were worse and they had to be improved, all this was discussed."

(Does this way of talking run in the profession? At his trial, Adolph Eichmann once apologized to the frustrated judge questioning him, "Official speech is my only language.")

Despite the colonel's thicket of verbiage, it was clear that at least part of his "political work" was to make sure that inmates fulfilled their quotas. Quotas were the bedrock of the Soviet command economy, the *gulag* included. Most prisoners had quotas for how much coal was to be mined, or timber cut, or grain harvested, every day. Unlike the civilian world, where Soviet managers routinely falsified such figures, quotas in the labor camps were a deadly serious business. Food rations were slashed for prisoners who didn't dig or cut or reap their quotas. This left them with still less strength to work, and once that cycle set in, it was usually fatal. Prisoners' memoirs are filled with accounts of frantically trying to trick the quota-counters by maneuvering yesterday's logs or stones onto today's pile.

Was this what the colonel—then much more junior in rank—looked for at his "meetings"? He did not say. Margarete Buber writes about inspection teams from Dolinka coming to the nearby Karlag camp where she was an inmate. "One evening all the occupants of the women's hut had to parade outside. A uniformed member [of the secret police] then read us out a list of sentences passed in Dolinka on various prisoners for offenses against the camp rules. There were two and three years extra for attempts to escape; two years for stealing skins; eighteen months for letting some sheep die; three years for the murder of a fellow prisoner; and seventy-five death sentences for repeated and willful refusal to work."

Colonel Volkov's account of making his rounds was far more vague and benign. When I pressed him for specifics, his language grew even more bureaucratic. "For example, I would be ordered to conduct two meetings in a certain department. People would disclose everything at these meetings, all their complaints, who was bothered by what. All these problems were solved, if not in one step, then by several steps, depending on the complexity of the issue."

As the colonel spoke into my tape recorder, his wife Olga bustled about the house. The coal miner son-in-law, whose shift had ended at midday, drifted in and out of the room. Suddenly the colonel's wife realized that a vanload of people was waiting outside.

A moment later, she herded the whole crew into the house. Since the living room was the room for receiving visitors, she sat everybody down on the couch and on chairs, all of which were lined up like soldiers, backs to the wall, directly facing the table where Nikolai Fyodorovich and I were talking to the colonel. It was like talking to somebody on stage. Arlie valiantly tried to gain me some privacy by asking to see the other rooms in the house, and by trying to engage the women in conversation in one of them. It was no use: the hostess determinedly shepherded everyone back to the couch and chairs that faced our table. Visitors must remain in the best room. Furthermore, visitors must be given tea— preparations for which began immediately.

The colonel's wife disappeared into the kitchen. Nikolai Fyodorovich and I continued asking questions of Colonel Volkov, under the eyes of the vanload of observers. I was sure their presence would make him clam up. Ludmilla Nikolaevna sat on the

couch, treated the interview as a public spectacle, and began translating, in her fashion, the colonel's words. "Lies!" she said loudly to
Arlie and Gabriel. "All what he says is lies!"

But the colonel knew no English, and didn't seem fazed. He
and his wife were clearly intrigued by the first Americans they had
ever seen "except in the movies." Also Nikolai Fyodorovich, in
gathering material for his notebook, had come to know the names
of Colonel Volkov's old colleagues. And even though he and the
colonel clearly distrusted each other, the colonel became visibly
more expansive as Nikolai Fyodorovich asked him if old Shirokov
was still around, and what ever happened to the Makatai camp,
and had the Molkovskoe department been under the command of
Zakhar Pavlovich, and didn't Colonel Maevsky used to live in this
very house? He did, said Colonel Volkov proudly.

Nikolai Fyodorovich seemed to be leading up to something
with all these names. Finally he said, "Among administrators and
guards there were masters of cruelty. Filimonov, for instance."

"They say every family has its black sheep," parried the colonel,
adding, with surprising frankness: "All the more so in work like
ours."

Colonel Volkov would have been ten years in uniform, and
seven in the Karlag system, when Khrushchev gave his famous
1956 "secret speech" denouncing the crimes of Stalin. What was
the impact of that here, I asked.

"It immediately stirred not just the employees of the system,"
he said, "but the whole country. How could all this happen? We'd
called Stalin 'our dear father,' and now suddenly it was clear what
he'd done. Nobody could understand it."

"You had believed everything!" interrupted Nikolai
Fyodorovich accusingly.

"How could I not?" protested the colonel. "There were documents from the Central Committee of the Party. The directions of
the city and regional Party committees—didn't I have to obey
them? Yes, I did. How could you dare avoid doing that? If there
was an Interior Ministry order, we had to follow it."

And so it went. The colonel's defense was, (1) he was just following orders; (2) the worst things happened before his day, in the thirties. ("We weren't here then!" butted in his wife vigilantly, as she served tea. "Such things did not happen in our time!"); and (3) he himself did not arrest people—that was done by others.

Most of this was literally true. Yet there was a ghostly vacancy in the picture of Karlag he drew. From the colonel's words you would not have known that it was a prison. Instead, he talked almost entirely about Karlag's role in the Soviet economy. He sounded like a proud regional Party boss. "We had our own agricultural experiment station. Cattle breeding was also advanced. A special breed of cow, Red Steppe, was raised here, also Kazakh Whiteheads. There were grain farms; we had poultry here. We sent food to the front during the war."

Similarly, when the colonel talked about what great scientists and physicians and teachers they had here in the good old days, he never mentioned that all these people were exiles or prisoners. His children had gone to the school whose ruins we had seen this morning. "God should send us more such teachers! There was Yuri Panarzhiev, a mathematician, a brilliant mathematician. He was a wonder! When you listened to him, you could not tear yourself away. Now there are no more teachers like we had before. Before, we never had a case when a young fellow leaving the school couldn't get into any institute [of higher education] in the country. There was never such a case. All of them went to Leningrad institutes, Moscow institutes, institutes anywhere."

The colonel said nothing about the twelve-hour work shifts for prisoners in the mines or the stone quarries. Instead, he got so carried away that he declared prisoners should have been grateful to have work. "Naturally, when a person works, he feels better! To serve a prison term without working is the harshest punishment."

How does someone like Colonel Volkov feel about *glasnost* and the withering away of the Communist Party? Not good, it was clear. Like all Russians who miss the old days, he was most upset by today's lack of discipline. And, as an old quota-enforcer, by the fact that no one is producing their quotas anymore. "In those days, only one word of Stalin's was needed, and everyone understood, from kids to the oldest people. Since we in uniform are used to discipline, it seems strange to us now when orders and laws aren't obeyed. Dear comrade, what have we come to?" He gestured despairingly. "What have we come to that we have nothing in the stores? Farms and factories don't fulfill their plans! Some even say that we should bring Stalin back to life for a couple of months—this is a joke, of course!—to bring order to the country. I say, let's find a Georgian who looks the same! Then people will be scared and will obey the laws. Discipline is undermined now, it's undermined everywhere."

Many houses in Dolinka are occupied by ex-prisoners and their descendants. I asked Colonel Volkov if he runs into these people on the street. It happens all the time, he said. "He'll say hello, and I'll stretch out my hand and say hello, and I've forgotten who he is. 'Oh, what, Mikhail Fyodorovich, you don't remember me? I worked there, and there, as such-and-such.' He'll tell me about his family, and I'll wish him good luck."

(In Vassily Aksyonov's novel *The Burn*, a former exile sees an ex-NKVD man: "I recognized him, but he didn't recognize me. They never do recognize us. There were so many of us, after all!")

After nearly two hours with the colonel, I began to run out of questions. So, I noticed, did Nikolai Fyodorovich, who started to make accusations instead. He spoke so angrily that I began to think there must be something in his background that we didn't know. "The system made us develop hostility to each other, the system itself!" he declared to the colonel. "Hostility, revenge, denunciations, spy mania, all that was encouraged—all stemming from the man himself, the man with the mustache. From top to bottom these base instincts were encouraged! There was an informer for every three or five people. It was terrible!"

Colonel Volkov remained civil enough on the surface, but I didn't know what feelings really lay in his narrow, wary eyes.

Just as I decided that it was definitely time for us to go, the colonel's wife, who had been out of sight for some time, reappeared with a folding table and tablecloth. There was no escape: we were in for a meal.

This was the last thing I wanted, especially with the tension starting to crackle in the air between the colonel and Nikolai Fyodorovich. But before we could leave, the food was already on the table. Lots of it: sausage, canned fish, fatty salami, lamb pilaf, bread, pickles, and a soup containing *kvass*—the old Russian peasant drink made from fermented grain. The colonel poured shot glasses of vodka for himself, me, and Nikolai Fyodorovich, although not for the women. He began proposing toasts to this first visit by an American family to Dolinka, and to peace and friendship between all peoples. He asked my patronymic, so he could toast me in the proper form. I felt even more apprehensive, having already seen too many endless Russian meals where the men drain glass after glass of vodka, mixed with lip-smacking praise of this elixir of life, while the women retreat to another room and worry about how to get them home.

Nikolai Fyodorovich seemed to feel as uncomfortable as I did. He, too, took only tiny sips at each toast, instead of downing the whole glass in one gulp, as the colonel did. After many toasts to us, the colonel was refilling glasses that were already full, or was missing them entirely when he poured. Finally the meal was over, and, despite our hosts' pleas to stay longer, we got up to leave.

But the vodka had made the colonel even more talkative. Before we left, he said, he wanted to show us something. From the bookshelf where the Shakespeare was, he pulled down a bundle wrapped in newspaper. It was his photographs. There were family photos and young soldier photos—group shots of the colonel and his secret police cadet classmates. And, amazingly, there were pictures of him at work. Although other aspects of his duties were doubtless less photogenic, these made clear what one part of "political work" was. In most of the pictures, the young officer Volkov, in uniform, was addressing an audience in a meeting hall.

Some listeners in the pictures were guards in uniform. But some were clearly prisoners—shaven-headed men with expressionless faces, in rough, baggy shirts. They were not emaciated, and sometimes several were even sitting up on stage behind Volkov himself. These were the trusties—common criminals who got extra rations and positions of authority over other prisoners. Like the *Kapos* in the Nazi concentration camps, they were a vital part of the *gulag* apparatus. Part of Colonel Volkov's job obviously had been to hand down the Party's correct line both to the staff of guards and to the upper, privileged caste of prisoners.

The colonel proudly volunteered a piece of information I had not known: that these trusties sometimes even worked as armed guards. "People were carefully selected for this purpose, worthy people. They were reliable. They signed special papers, they were taught how to handle weapons. They were positioned at the watchtowers and they were guarding . . . themselves!"

On our way out, the colonel showed us his orchard and garden, and hutches for the three dozen rabbits he is raising. Meanwhile, his wife was whispering to Ludmilla Nikolaevna (she told us later) the only question either she or her husband asked about us, the first Americans they ever saw: "How much money do they make?"

At last we were on our way, past the ruined prison again, and back across the flat, open steppe toward Karaganda. We all were happy to be out of the colonel's house, and a shared sense of relief pervaded our oddly assorted vanload. Nikolai Fyodorovich began declaiming from his notebook again, but Arlie interrupted and asked him, with Ludmilla Nikolaevna translating, "What did you think of the colonel?"

"He knows much but says little," said Nikolai Fyodorovich, "And ninety percent of what he says, you shouldn't believe."

"Have you forgiven him?" Arlie asked.

"Nyet!" said Nikolai Fyodorovich. "But he used to be powerful and I used to be low, and now we're equals."

Although he still wore his black homburg and black trench coat, he suddenly seemed more human. Arlie asked him more, and for the first time he told us something not in his notebook: that he himself had arrived in Karaganda at the age of four because his parents were *kulaks* thrown off their land and "into the snow" during the collectivization of agriculture. After their arrival, his mother was put in prison, "for taking a few ears of grain.

"For seventy years they deceived us," said Nikolai Fyodorovich, in a weary, resigned tone. "And they made us deceive each other"—referring to his time as a history teacher. "Now at last we can get to the truth." He tapped his notebook, but with more sadness than anger.

The day's trip left me strangely dissatisfied. I had neglected, I realized, to ask some obvious questions. In Colonel Volkov's household, who was it who loved Vysotsky and read Shakespeare and the thousands of other books on the shelves? The colonel, his wife, and the coal miner son-in-law aimlessly wandering about the house seemed unlikely candidates. That left the Volkovs' daughter, who was not there. Did this mean she had gotten her love of literature from a prisoner-teacher at her old school?

I think I also felt dissatisfied because the people we met in Karaganda were not playing the parts I had unconsciously assigned to them. The story was more complicated. The colonel, with his rabbits and fruit trees and farmer's hands, didn't completely fit the role of villain. Back in Moscow, Olga Infland, describing how she had wept at Stalin's death, didn't fit the role of noble victim. Nikolai Fyodorovich, with his black hat and prepared spiel, didn't fit the role of hero, grappling boldly with the dark legacy of the past. Ludmilla Nikolaevna, although we warmed to her thoroughly, belonged in a farce, not a tragic drama.

In a study of the testimony of Holocaust survivors, Lawrence L. Langer points out that these men and women usually tell stories far more despairing than their interviewers want to hear. Langer looks at videotapes and transcripts, showing where the interviewers interrupt their subjects, or ignore things the survivors have said. The interviewers, collecting life histories for a video archive at Yale, seem unconsciously trying to elicit the same story as scriptwriters and novelists usually tell about the Holocaust—a terrible story to be sure, but one that ends with the day of liberation from camp and with the triumph of the human spirit. But, Langer points out, camp inmates were stripped of all dignity; the day of liberation was often one of numb bewilderment; many survivors have attempted suicide; the human spirit did *not* triumph.

One must be equally wary of trying to turn the experience of Stalinism into a morality play. As I look back now on the little slice of that experience I saw in Karaganda, what it suggests is more complicated than lessons about good and evil. It was a lesson, above all, in our capacity for denial.

This denial appears in everyone, in different ways. Colonel Volkov carried out orders for years, cheerfully convinced, I imagine, that he was building a better future. He claimed to be shocked by the revelations of Stalin's crimes in Khrushchev's "secret speech." On this point, I believe him. In the years before then, how could he have summoned up the energy to enforce the quotas, to rise in the bureaucracy, to have pictures taken of his speechmaking, if he did not have faith that he was doing something worthy? Unless these laggard prisoners are whipped into line, we won't fulfill the Plan. Few evildoers, at the time, believe they are doing evil. What the colonel had to deny was any intuition that the thousands of people he was guarding and punishing were not dangerous spies deserving the harshest of treatment. His trainers, his superiors, his peers, all helped him deny it by denying it, too: the number of known defectors from the Soviet secret police over the years is minuscule.

Olga Infland's weeping at Stalin's death, the denial of evil in a different way, seems more unexpected. But she is not the only ex-prisoner I have heard speak of such tears—and usually with apologetic embarrassment, as if they still could not comprehend today why they had grieved then. The writer Olga Berggolts wrote a poem of sorrow and admiration on Stalin's death, even though her husband had died in prison and she herself had been jailed while pregnant and was beaten so badly that she lost the baby. There is an echo of something ancient in these women's grief. During the reign of Ivan the Terrible, Prince Repnin, impaled on a stake, praised the Tsar, his executioner. And when Ivan himself lay dying in the Kremlin, writes the Marquis de Custine, the great French observer of Russia, he was "mourned by the entire nation, not excluding the children of his victims . . . the insoluble problem for the philosopher, and an endless source of astonishment and awesome contemplation, is the effect that this unequaled tyranny produced on the nation decimated by it. Not only did the people not revolt, its loyalty was actually increased."

In her memoirs of the Stalin years, Nadezhda Mandelstam blames people's denial of reality on a naive Utopianism: "In the pre-revolutionary era there had already been this craving for an all-embracing idea which would explain everything in the world and bring about universal harmony at one go. That is why people so willingly closed their eyes and followed their leader. . . ." This is the classic explanation, and it does say much about why so many early Western visitors to the Soviet Union saw nothing but folk dancers, gymnasts on parade, and happy schoolchildren. When reality didn't fit the formula, they believed the formula.

By the late 1930s, however, Russians like Olga Infland had reached a second generation of denial: while they might admit that Utopia had not yet worked out as advertised, they failed to see that the murderous arrest machinery devouring everyone in sight would harm *them*. This kind of denial has more to do with a deeper, more general human reluctance to acknowledge any catastrophic threat. Most of the people who boarded the trains for Auschwitz believed they were heading for "new resettlement areas in the East." Looking back at them today, we are incredulous, just as we are in looking at Great Purge victims: *why couldn't they see it coming?* Yet is this blindness really so different from that of a person of our own time and place who refuses to believe that he or she, or a loved one, is fatally ill? Or, for that matter, is it so different from our various collective denials, such as the assumption that we can keep on pouring pollutants into the Earth's air, soil, and water without someday suffering the results?

Especially when it involves some horror beyond one's experience, denial comes in several layers. You can know the truth, yet at the same time not accept it. The historian Walter Laqueur tells the story of Jan Karski, a brave Polish Resistance fighter who smuggled himself into, and out of, a Nazi concentration camp in 1942. He then made his way to the Allied capitals, to tell the world what he had seen. In Washington, he met with Supreme Court Justice Felix Frankfurter. Frankfurter was horrified by what Karski reported, but said he couldn't believe him. Karski objected, and then Frankfurter explained what he meant. He knew Karski was telling the truth, he said. He just couldn't believe him. That, I suspect, was how Olga Infland and her mother heard the lawyer's warning that they should flee to avoid arrest.

Nikolai Fyodorovich in Karaganda was also denying something, and in some ways his denial was surprising. "For seventy years they deceived us. . . ." But how could he let anyone deceive him when he had arrived in Karaganda in the first place because his parents were torn away from their land, and when his own mother had gone to prison for stealing "a few ears of grain"? He had to deny the life experience of his entire family. In the end his tragedy—and his resigned tone suggested that he was aware of this—was not that "they deceived us" but that he had deceived himself.

In the mid-1950s, Solzhenitsyn estimates, more than two thirds of Karaganda's population were exiles like himself, who had to report to the police at fixed intervals. "The largest of [the] provincial capitals of the Archipelago was Karaganda. It was created and filled with exiles and former prisoners to such a degree that a veteran *zek* [from *zakliuchoni,* or prisoner] could not walk the street without running into old acquaintances. The city had several camp administrations, and individual camps were scattered all around it like the sands of the sea."

And yet it was possible for people to be born and to grow up on this sea, and still be unaware that it lay beside them—a kind of denial so deep and unconscious that we almost need another word for it. This denial by bystanders permits the whole apparatus to run without dissent, without disruption, without even being noticed. We met one such bystander our last evening in Karaganda.

Ludmilla Nikolaevna, the English teacher, invited us to dinner at the home of her cousin Yulya, a professor of literature at the local university. Yulya was a spirited, cheerful woman of fifty-seven. She told us that she grew up in Karaganda, but without realizing what went on here. She said this not as an excuse, or in repentance, but as a simple statement of fact, and of a fact that now clearly troubled her. The father of her best friend, who lived next door, was a colonel, a good family man, kind to his wife and daughter. "But he always went to work very late—the driver always came for him at night. Why? I wonder now. Interrogators work at night. Is that what he did?" Yulya sometimes went to the theater with the colonel's daughter, she told us. She never realized then what she knows now: that all the performers were prisoners.

I would like to think that had I been a bystander at that place and time, I would not have denied the evidence, I would have boldly seen the fear, the camps, the terrible injustice. But would I? Growing up, we take our clues from those around us; if our families, our teachers, our newspapers don't notice something, we are not likely to notice, either. Only when we suddenly see the familiar world from a new angle do we realize that something is amiss. For Yulya, this did not happen until she was sixteen and went away to continue her education in Moscow. Only then, for the first time, did she begin to understand things. What made her aware, she said, was the look that came into people's eyes when she named the city she came from, Karaganda.

6
Souls
Living
and Dead

Among the various types of denial, that of the victims, like the prisoners who wept at Stalin's death, may be the most puzzling. But it is the denial of jailors and executioners, their ability to not see their victims as members of the same human family, that is the foundation of any police state. Without men like Colonel Volkov willing to carry out orders, the system would not have functioned, and 20 million people would have lived.

Where do you begin, if you need to deny that someone is a human being? First, you call them animals. In the Moscow show trials, prosecutor Andrei Vyshinsky called the defendants "vermin" and always demanded of the court, "Shoot the mad dogs!" The official emblem of the NKVD was a sword striking down a serpent. One poster from the 1930s gives Trotsky paws like a dog; another, mixing species, shows Russian Orthodox clergymen with hoofs, horns, and claws. A cartoon from 1937 has the heads of Trotsky, Bukharin, and various others emerging from the body of a snake; the snake is being strangled by NKVD chief Nikolai Yezhov, who wears a steel glove. A cartoon from the next year shows some of the same "enemies of the people" with pigs' bodies, wallowing in a food trough. This is the key to denial: your enemies are not people, but rabid dogs, rats, pigs, vipers.

This habit of thinking gradually spreads from executioner to bystander. It must, for without separating those before you into us and them, you cannot remain peaceful and untroubled in a society where imaginary enemies are shipped to the *gulag* or Jews are sent to the gas chambers, or the poor starve, or the homeless lie on sidewalks.

On this same Siberian trip that took us to Karaganda, we also visited Novosibirsk, which lies some 500 miles beyond, to the east. Among the people I met there was a man who, like Yulya in Karaganda, had grown up surrounded by prisoners, NKVD officers, and labor camps. But, unlike her, he could not deny their existence; it was too obvious. The striking thing was how everyday this whole world seemed. It even included a local executioner.

Vladimir Nakoryakov was fifty-five years old, a tall, forceful, burly man, director of a prestigious physics research institute just outside Novosibirsk. Among executives anywhere, office size is usually proportional to rank, but those at the top of any Soviet hierarchy traditionally have had an additional badge of status—a long conference table set at right angles against a desk. It suggests, From my position I can preside over a whole room full of underlings. The conference table in Nakoryakov's office was long enough to seat twenty people. For our talk, though, he sat down across from me at the far end of the table, a room's length away from his desk, as if to say that he was putting aside his high position for this conversation.

"I grew up as the son of an enemy of the people," Nakoryakov began. "It was a most unusual life." In 1937, when Nakoryakov was a small boy, his father, an Old Bolshevik and Red Army veteran, was shot. The NKVD had charged him, for reasons no one could discern, with being a spy for Argentina. After his execution, Nakoryakov and his mother moved to eastern Siberia to live with her father, who had been a pre-revolutionary businessman, educated in Belgium. "He had been a company president, living abroad. When the Revolution began, he had returned to fight for the Tsar. This man who had been educated abroad, had been a capitalist, had defended the Tsar—he was to die naturally in 1952. Whereas the man who fought for Lenin, was a hero of the Civil War, was secretary of the Party regional committee—he was shot.

"We lived in a wooden one-story house. It was just a small cottage, probably ten meters by seven. Grandfather was a wonderful man. He knew English, French, and German, living in this God-forsaken place. Half the house was occupied by Grandfather and us, and the other half by an NKVD commander. He was said to be a very cruel person, and many people died because of him—but Grandfather, a Tsarist army officer, kept right on living in the other half of the house. The two men were friends. They loved to play chess. Probably the NKVD boss didn't arrest Grandfather because he didn't want to lose a chess partner."

At home and at school in this small Siberian town, nobody ever mentioned the name of Nakoryakov's executed father. Yet this was not so unusual. "In those years [the 1940s] nobody asked questions, because nobody had fathers. Either the fathers had gone to the war and been killed, or they had been arrested. In my class at school, of thirty kids, only two had fathers."

As he spoke in brisk, no-nonsense tones, Nakoryakov constantly wrote or sketched on small white file cards: a floor plan of his childhood house, any date or number he mentioned, and now a rough map of Siberia to show me where this town was—on the Trans-Siberian Railroad, near the Pacific coast.

This region was honeycombed with labor camps, and many officials of the camp system lived in the town. One of their neighbors, Nakoryakov said, "was an executioner. He would come home, and tell stories to children of how he'd shot people. I remember him and the local militia chief having target practice, and his saying that it was more interesting to shoot when your target is a running man: 'There's this special feeling you get.' I was young then. One's memory is selective. But the impact of his words was so strong that I remember them. It's a vivid image in my mind: this man is saying *that* as they are aiming their pistols at the target." (Again, an echo of animal imagery: what is shooting a running man like, if not hunting?)

Nakoryakov's childhood world was one of executioners, labor camps, vanished fathers. The Trans-Siberian Railroad ran through his town. And since it led to the Pacific ports from which freighters sailed north to Kolyma, a major cargo on the trains was prisoners. "All the convicts were going to Kolyma. There were certain hours at the railway station when the convoys of convicts moved through, and everyone had to leave the station. The tracks are here"— Nakoryakov rapidly sketched a diagram on a white card—"the platform is here. When the trains were moving to the east, everything was guarded by the militia and nobody was allowed inside the fence. But there was a volleyball court behind the fence, and we always managed to get onto it. Or else we were there and didn't leave. That's why, when those trains were coming through, I could watch them. Stuffed with people, behind bars. Their faces. You see, we were so accustomed to the fact that convicts existed. There were soldiers, and there were convicts.

"This was the life that surrounded us. We didn't think another life could exist. Everything was so ideologized, so politicized. I remember one case. A schoolteacher stabbed his wife because he was jealous. In the morning they brought together all the students, and the principal of the school explained things this way: 'The world situation is very difficult: imperialists are on the move everywhere and there may be a war with the United States. *That's* why this man killed his wife—to get into a labor camp so he could avoid the draft. . . .' "

Living in that world, did Nakoryakov have to deny anything? Only, perhaps, the idea that the whole structure was not perfectly normal. Today, however, he is obsessed with it. "How many guards worked in the labor camp system? More than one million. I've calculated it." He restlessly scribbled numbers on a white card, adding, subtracting, explaining his reckoning. "The procedure for arrests was the same as for steel production. The NKVD was a regular state department, which had a quota of a certain number of people to arrest. A certain amount of steel was needed, while this department agreed to convict another twenty thousand saboteurs. If they arrested fewer, it might mean that some NKVD employees would be out of work. That's why they wanted to increase the number of people arrested, to maintain their structure. The more they arrested, the better they lived. It seems to me that the moment this organization came to life, we were all doomed."

What does it take to enjoy shooting a running prisoner? Surely it helps if you feel you are shooting at something less than human, a mad dog. But denying someone's humanness—whether you pull the trigger or merely sign the order—must be preceded by other denials: of right and wrong as values that transcend obeying orders, and of the idea that Party doctrine of the moment is not an infallible guide. It was those denials and that sense of infallibility that led so many early Bolsheviks down the long path toward becoming executioners, or victims, or, quite often, both.

There was another person I wanted to meet in Novosibirsk who might have something to say about all this. And who I thought, might even be an example of some of these problems himself, for he came from the world of those at the very center of the Russian Revolution.

When the Bolsheviks seized power in 1917, the new Soviet Union's first official head of state was a dignified-looking man with a professorial bearing, Lev Kamenev. The name "Kamenev" is an adopted, Party name; it means "man of stone." Kamenev was born Lev Rosenfeld. During the decade before World War I, he worked with Lenin and the other Bolsheviks who were then in exile in Western Europe. The year 1914 found him back in Russia. The Tsar's government soon banished him to a remote village in Siberia. A 1915 photograph shows him with a group of fellow Bolsheviks confined there. Kamenev appears very much the serious intellectual, in a mustache and beard. Standing next to him is a surprisingly benign-seeming young man with the look of a country bumpkin, in a broad-brimmed hat: it is the young Joseph Dzhugashvili, later known as Stalin, or "man of steel." In the end, the man of steel got the better of the man of stone. In copies of the photograph published in the 1930s, Kamenev's figure has disappeared, replaced by some shrubbery in the background.

Kamenev and Stalin found themselves together in another desolate Siberian town two years later, when the uprising of March 1917 toppled the Tsar from his throne and installed a Provisional Government. When they heard the news, the two men set off together for the long train journey to Petrograd. Both were present at the secret, ten-hour meeting in a Petrograd apartment to plan the Bolshevik seizure of power of November 1917. Before the gathering broke up at 3:00 A.M., both were elected members of the Party's inner circle, the seven-man Politburo. Immediately after the Bolshevik coup, it was Kamenev who announced to the world the members of the new government.

Kamenev remained on the top rungs of Soviet power for the next seven or eight years. He ran the Party organization in Moscow. He went to Western Europe in 1918 to rally leftists there, for the Bolsheviks were hoping for other revolutions to erupt in the wake of World War I. But he was expelled by Britain, and came back to Petrograd to report, "Comrades, we are alone."

An able conciliator, Kamenev was always the person called upon to mend rifts in the Party, or to negotiate with other groups. One colleague wrote that Kamenev's attitude toward his political enemies was "so gentle that it seemed as though he himself were ashamed of the irreconcilableness of his position." Kamenev believed the Bolsheviks should form a coalition government with the other left-wing parties—a major heresy, for it would have meant, in effect, majority rule. But in this, as in many other things, he first lost out, then bent to the Party line.

One of the great psychological puzzles of the years after the Revolution is why so many intelligent, educated Communist Party members did just this. They looked into the abyss, saw the arrests, the midnight executions, the suppression of all dissent first outside and then inside the Party, recalled the reign of the guillotine that had followed the French Revolution, vainly spoke out in protest—and then came back loyally to the Party fold. The analogy of communism as a religion only partly explains their actions. For surely no religion on earth has ever seen so many vocal heretics return to the church. For these wavering Old Bolsheviks, who had been in the Party since before the Revolution, there was no possibility of salvation outside the faith: you were either on one side or the other, with the forces of the future or the forces of the past. There was no space in between. It was this that forced them to deny so many ominous signs of the new order emerging around them. The tragedy of these men and women was not only that almost all of them were shot, but also that by their return to the Party and by their denials they were complicit in their own deaths.

Both before and after the Revolution, Kamenev wrote prolifically, and he edited the works of the great nineteenth-century liberal Alexander Herzen. But whatever his intellectual bent or his personal reputation for conciliation, he shared that near-mystical belief that the Party was the sole embodiment of correct thinking. And he shared, also, the desire to root out any thinking held to be incorrect. Kamenev was present, for example, at the 1922 night interrogation of the distinguished philosopher Berdyaev. Berdyaev's Christian Socialism was not to the Bolsheviks' liking, and he and a number of other non-Communist intellectuals were sent off into European exile on what later became known as the Philosophers' Ship. It was a mild punishment by later standards, but a fateful symbolic step on the road to the Great Purge.

In the mid-twenties, Kamenev and his long-time collaborator Grigory Zinoviev joined with Stalin in a "triumvirate" of Politburo members that for several years in effect ran the country. This was before Stalin had seized all the levers of power for himself, and many top officials openly distrusted him. At one point, Kamenev even implored the Central Committee not to fire Stalin as General Secretary of the Party. Like so many others, Kamenev had no sense that his beloved Revolution could give birth to a new Tsar.

Once Stalin had consolidated his power base in the late twenties, he then swiftly packed the Politburo and Central Committee with more compliant henchmen. This meant deposing all the well-known Old Bolsheviks like Kamenev and Zinoviev. Stalin sent

Kamenev off as Ambassador to Italy—an insulting slap, for it meant being envoy to Mussolini.

This began Kamenev's long slide toward disgrace and death. It was hastened all the way by his continuing faith that the Party was endowed with theological rightness, even when his human instincts must have told him it was catastrophically wrong. For Kamenev's generation, there was also a secular side to their professions of faith, for being in the Party's good graces was a prerequisite for holding any power. Eventually, it became a prerequisite for staying alive.

In the last decade of his life, as Stalin toyed with him, Kamenev was three times expelled from the Communist Party, and three times asked for reinstatement and forgiveness. After a temporary alliance with Trotsky failed to oust Stalin, Kamenev apologized, and was allowed to rejoin the Party in 1927. "We are surrendering," he told a Party Congress, "because we are deeply convinced that a correct Leninist policy can be made to triumph only in and through our Party, not outside it or despite it."

Stalin never permitted Kamenev to return to the ranks of leadership, however. Some five years later, he was again expelled from office and sent to Siberia. In 1933, he recanted once more, and was allowed to return to a minor job in a publishing house. At the Party Congress of 1934, Kamenev repented of his heresy with a consciously religious metaphor:

"I want to state from this rostrum that the Kamenev, who from 1925 to 1933 struggled with the Party and its leadership, I regard as a political corpse. I want to go forward without dragging the old skin after me, if you'll excuse the Biblical expression. Long live our, *our* leader and commander Comrade Stalin!"

Kamenev's one-time ally Leon Trotsky could be arrogant and pitiless, but he never voiced the endless recantations of so many of his former comrades. In his exile, hounded by Stalin's agents, Trotsky wrote presciently: "The future historian, who will wish to show how ruthlessly an epoch of great upheavals devastates characters, will take Zinoviev and Kamenev as his example. . . . Like Gogol's hero, Stalin collects dead souls for the lack of living ones."

Although Kamenev was barely fifty at this time, a photograph of him with his wife and two sons shows his hair and beard turned completely white. His wife protectively hugs their younger son, who is perhaps four or five years old, as if she knows something is going to happen to them all.

In 1934 Kamenev was again arrested, and sentenced to prison. Two years later, still in custody, he was charged with being part of a vast conspiracy, including the Gestapo, Trotsky, and various other forces, that aimed to assassinate a whole string of top Soviet figures. This was to be the first big show trial of the Great Purge. But beforehand, Stalin wanted to make sure that the key defendants would give the necessary confessions on the witness stand. To soften up the prisoners, he had the NKVD put the heat on—literally, by turning on the heat in their cells in the middle of a hot Moscow summer. After Kamenev and Zinoviev repeatedly wrote him from jail, frantically pleading for mercy, Stalin summoned them.

Stalin's biographer Dmitri Volkogonov reports:

"Stalin did not respond to their greetings nor invite them to sit down. Pacing up and down, he offered them a deal. . . . If they would confess to everything . . . he would save their lives, or rather, he would try to save them. And then he would do his best to get them released."

Desperate, Kamenev and Zinoviev agreed.

The trial opened, and for the first time a shocked international audience saw the spectacle of high-ranking Old Bolsheviks, the makers of the Russian Revolution, pleading guilty to taking part in conspiracies that were not only absurd but impossible—Kamenev had been in prison or exile for most of the period of the alleged plot.

"We have served fascism," Kamenev declared when it was his time in the dock. "Such has been the path we took, and such is the pit of contemptible treachery into which we have fallen." Apparently still holding onto the hope that Stalin would follow through on the promise to save him, Kamenev virtually called for his own execution: "Twice my life was spared, but there is a limit to everything, there is a limit to the magnanimity of the proletariat, and that limit we have reached."

"Stalin doesn't have to cut heads off," observed the poet Osip Mandelstam. "They fly off by themselves like dandelions."

Stalin's genius lay in getting others to do his work for him—not only his subordinates but his victims. Brave souls who refused to confess to anything were executed in secret; the show trials were for those who had agreed to plead guilty. The sight of people of Kamenev's stature accusing themselves of treason helped convince many Soviets and foreigners that even if the entire conspiracy charge wasn't literally true, well, still, there must be something to it. Diplomats, foreign correspondents, visiting members of Parliament and the like filled the elegant columned hall used as a court room, and many sent home reports saying justice was done.

The prisoners were sentenced to death, and officials immediately announced that the sentences had been carried out. After this, Kamenev and Zinoviev were returned to prison and paraded before the next batch of prisoners, living proof that Stalin had kept his promise to spare their lives.

Then they were taken down to the basement and shot.

Kamenev "was not killed by the first shot," writes Robert Conquest, the preeminent Western historian of the Purge, "and the NKVD lieutenant in charge became hysterical and kicked the executioner with a cry of 'finish him off.'"

Why did so many Old Bolsheviks confess to preposterous crimes? To the usual answers of torture, and belief in following the demands of the Party, in Kamenev's case we must add one more: Stalin's vague promise of salvation—and Kamenev's obvious hope that even if he couldn't save himself, he might at least save the lives of his family. This was clear in his final speech to the court.

"When Kamenev had finished his plea," writes Conquest, "and had already sat down, he rose again and said that he would like to say something to his two children, whom he had no other means of addressing. One was an Air Force pilot; the other, a boy. Kamenev said that he wanted to tell them, 'No matter what my sentence will

be, I in advance consider it just. Don't look back. Go forward. Together with Soviet people, follow Stalin.' He then sat down again
and rested his face in his hands. Others present were shaken, and
even the judges are said to have lost their stony expressions for an
instant."

But these words did not help. Stalin's ferocity toward the Old
Bolsheviks knew no bounds; it was like the clan warfare of his native Caucasus. In the months after the trial, the NKVD seized almost everyone named Kamenev whom they could find. And, most
people long believed, they killed Lev Kamenev's entire extended
family. According to Volkogonov's authoritative recent biography
of Stalin, "Kamenev's wife, his two sons (one of them still a minor),
his brother and his brother's wife, all perished."

But Volkogonov is wrong.

Kamenev's younger son survived. And one reason I went to
Novosibirsk was to find him.

Kamenev's son survived because he changed his name. He had
been seven years old when the chief Purge prosecutor Andrei
Vyshinsky had screamed that his father must be shot like a mad
dog. The father had been steeped in classic Bolshevik dogmatism
and there were signs that the son, Vladimir Glebov—he now uses
his mother's last name—might be as well. One of the few things I
knew about Glebov was that he was a teacher of philosophy, which
throughout Soviet rule generally meant Party-line Marxism-
Leninism. And, though Glebov was a labor camp veteran himself,
I had heard that he was not in Memorial and belonged instead to
some other organization of survivors. This sounded too much like
an echo of the sectarian struggles of the old days. I expected to find
a present-day version of an austere, stern, unbending revolutionary,
denying any view of reality save his own.

In the courtyard of a battered, gray Novosibirsk apartment
building, the last of the winter's snow is melting into mud. The
apartment I'm looking for is on the fifth floor. The slight, lithe
man who opens the door is in his early sixties. His top half looks

remarkably like Lenin: goatee, mustache, a balding skull, a shirt and necktie. But from the waist down, everything is distinctly un-Leninist. Vladimir Glebov is wearing Bermuda shorts, bright blue slippers with red trim, and no socks. And he is laughing—an infectious, welcoming gale of laughter.

His apartment is shared by a friendly cat and, packed onto shelves and in huge piles on the floor, what looks like about ten thousand books. Most unexpected of all, one heap of books is by, or about, Emily Dickinson.

"She is the best *muzhchina* [the word for man, as opposed to woman] of your poets!" he says with enormous enthusiasm. Like many Russian intellectuals, Glebov reads English fluently but seems unwilling to speak it, having had few chances to practice.

He just moved into this apartment several days ago, he says; that's why so many books are still in piles. He is building more shelves for them. Thanks to his time in Stalin's camps, he explains gleefully, he knows carpentry, "so if they retire me from teaching, I'll have a trade."

Although he teaches philosophy, Glebov says his real academic love is Slavic history and culture; his specialty is nineteenth-century Serbia. Why? When he was a university student in Stalin's last years, the slightest departure from orthodoxy could be fatal. "Our shrewd professor told us an aphorism: Here in Russia, a self-respecting historian must study something not closer than a hundred years to the time of his birth, and not closer than a thousand kilometers to Moscow. Only then can you be an honest historian!"

He erupts in another contagious burst of laughter, and leaps up to find a monograph he wrote on Slavic ethnography. Something seems almost catlike about his movements as he hops nimbly around the apartment, fetching tea, books, and knickknacks.

"I was born in 1929, between two periods of exile of my parents. My father was jailed three times under the Tsar, plus twice in England, once in Finland, and three times under Stalin. My mother was in jail once under the Tsar and later three times under Stalin.

"My first exile was when I was four. I didn't know it at the time! When I was arrested in 1950, they told me that I was an old offender, and that now I was in for my third term. 'I know about one other,' I said, 'but what was the first?'

" 'Well,' they said, 'we have it recorded right here: 1933–37, exile.'

"I said, 'I'm sorry, but I was four years old then!'

"Then, a year and a half ago, our KGB gave me a call. They said, 'Vladimir Lvovich! We've been sorting through old archives, and came across copies of letters between your mother in exile and your father in prison. Do you want them?' Those letters were written in 1935 and they called me at the end of 1989. At that moment, I realized, dear comrades, that such an organization can actually be useful! How otherwise would those letters from half a century ago have been preserved?"

Some of the human qualities of Lev Kamenev, Glebov's father, seem to have survived all the long years of doctrinal disputes and recantations; in his letters to his wife and young son, Kamenev tried to tell funny stories from prison and exile that would amuse a small boy. One was about a hedgehog that went to sleep inside Zinoviev's shoe.

Kamenev was shot in 1936. Soon afterward his wife, Glebov's mother, was arrested, "and then my own adventures began." For some years, Glebov did not know exactly when and where his mother met her death. Only a year ago was he able to learn that she was shot in 1937. After her arrest, Glebov was briefly taken in by a woman neighbor in the town where he and his mother had been exiled. Then, for the next nine years, "I was wandering through Siberian orphanages."

The first orphanage was in the subarctic timber country of western Siberia, "the same latitude as your Alaska." Just getting there required a long trip by river barge, then two weeks' trek through the forest, in a horse-drawn cart. About 40 percent of the orphanage children were offspring, like him, of Old Bolsheviks, purged Red Army officers, and other "enemies of the people." Another 20 percent were Gypsies: "A year before, they had arrested an entire Gypsy tribe for roaming, for being nomadic. They were put into trains and a barge and brought to the forest in October. Over the winter they died. When the adults were frozen to death, the kids were sent to the orphanage.

"Another 40 percent were juvenile delinquents. My buddy, who was nine or ten, had been a *fortochnik*. A gang of thieves would have a ground-floor apartment as their target. At night, when it was hot, the *fortochka* [a small, high pane in a casement window] would be open. The adults would lift him through the *fortochka;* his job was to go through the apartment and unlock the door. He would have a Finnish knife between his teeth, and if somebody woke up, he would stab him—professionally, just with one stroke—so this person wouldn't scream and wake up everybody else.

"We saw him being questioned by our supervisor: 'Ivanov, how many people did you kill?'

"He said, 'Damn! Seven? Or eight? I didn't count them.'

"My buddy taught me some things, which helped me a lot in my later life, about protecting myself. Here I have one scar, and here another. . . . When people are attacking you with a knife, you need to know how to fight back. The main principle is to respond in advance, not to let them hit you. That was our happy Soviet childhood!"

Where along the line did he change his name?

"Another Russian joke! I don't know who did it, or whether the people who did it were saving their own lives, wanting to not attract the NKVD's attention, or whether they were saving me. Some supervisor at the orphanage did it. I came there as Kamenev. About a year later I was summoned and they informed me, 'You're going to be Kamenev-Glebov.' Another year passed, and the Kamenev part disappeared and I became Glebov.

In 1945, the sixteen-year-old Glebov was allowed to go to Leningrad to study. A secret police officer at the university called him in and gave him some advice: "You can study only because we've forgotten about you. Don't remind anybody of your existence."

Glebov tried his best to keep a low profile, sticking to the nineteenth-century Balkans, suspecting that his luck would not last. It didn't. "I shared a dormitory room with an informer. We were three young men and he was the fourth. One day: Knock-knock! The superintendent came in, and three guys in plain-clothes. I thought, 'This is the fourth year I've been waiting for you fellows to come!'

"They said, 'Pack up! Your studies are over.' I knew that as long as Stalin was alive, I was supposed to be in jail. I had been free by mistake. Usually, people like me they didn't let go. Their exile term would be over, and they'd be sent to a camp; their camp term would be over, and they'd be exiled; and so forth. I knew that sooner or later they'd remember me. Now they had remembered."

At his interrogation, Glebov did not try to maintain his innocence, he says; he just told the interrogator, "Listen, captain. I'll be in jail as long as *He* is alive. But I'm twenty and He is seventy. . . ."

There remained the question of exactly what Glebov was to be accused of, since his real crime was being the son of one of Stalin's long-dead rivals. "They kept me there for some nine months before they found what to charge me with." He was finally found guilty of being an "aesthete."

Another burst of infectious laughter. "You know, I'm probably the only person in the world with a certificate, validated with an official stamp, that I'm an aesthete. Oscar Wilde didn't have it! Thoreau didn't have it! But I have this indictment which reads, 'Aesthetic approach to literature.' As a student, I had said that I could not stand Mayakovsky [an officially approved poet]; I loved Blok, Marlowe, Emily Dickinson. So they found somebody who testified that I'd said all this." Another witness testified that Glebov had also once called Stalin a dictator.

He was sentenced to ten years. "For slander of the leader of the world's proletariat and for an aesthetic approach to literature— only ten years! I was sent to work on the Volga-Don Canal. I worked in a sawmill. In Nizhni Novgorod, I spent some fourteen months in a prison built in Catherine the Great's time."

Finally, in 1956, after nearly seven years had passed and Stalin was dead, one of the investigative commissions Khrushchev sent around the country's prison system reached the camp Glebov was

Alone of the Russians I've talked to so far, Glebov is not afraid to dream in the visionary terms of social transformation that so inspired the radicals of the last century. But his is a gentler, more modest and tolerant vision, not the rigid, one-size-fits-all Utopia of his father's generation. It is exhilarating to hear him talk, a reminder of the idealistic strand of the Utopian impulse, which in Russia was so overwhelmed by a darker, bloodier side. Glebov's own dream is the one—so easy to imagine, so seldom tried—of finding some way to combine socialism and democracy. He wants a socialism that honors human rights and free speech, and embodies "a clever, flexible use of your experience" in the West.

But, Glebov cautions, he is not messianic about his vision, like the dogmatists of old. "This is for Russia, not for the whole world. And we shouldn't claim that our path is the only one and that those who disagree are swindlers or cosmopolitans or"—he smiles—"aesthetes. . . ."

Glebov has three grown children and several grandchildren. At the end of our long afternoon of talk, his son stops by to see how his father is settling into the new apartment. We move to the kitchen and all share a snack. With a great flourish, Glebov produces a small, whole smoked fish from a cupboard, and deftly bones and slices it with a knife, cheerfully tossing extra bits to the cat.

Only after I've left does it occur to me that Glebov has more reason to be bitter and cynical than anyone I've met in Russia: his father shot, his mother shot, his brother shot, a decade in orphanages, and seven years in the *gulag*—which, now that I think of it, he barely even bothered to talk about. But no one has made me laugh so much in months. Each chortle from this odd-looking figure in his Bermuda shorts and sockless slippers is another sign of his triumph. From the orphanage friend who taught him self-defense, he learned his lesson well. At every stage, Glebov has parried the system's blows before it could wound him. And his main weapon is laughter. I can still hear the echo of his final joke.

Stalin telephones secret police chief Beria and says, "Lavrenti, help! I've lost my pipe!"

Beria says, "Don't worry, Comrade Stalin, I'll get on the case right away."

The next week the maid is sweeping the floor of Stalin's study, and finds the pipe.

Stalin telephones Beria again: "Everything's okay, Lavrenti. I found the pipe."

Beria says, "But Comrade Stalin, I've arrested ten thousand people and charged them with stealing it!"

Stalin says, "Shouldn't we let them go?"

Beria says, "But they all confessed!"

Stalin says, "Oh. Well, in that case, let's give them all ten years. . . ."

Ukraine famine victims, 1933.

7
Believers

As time went by and I talked about the after-effects of the Stalin era with more and more people—in Moscow, on our trip to Siberia, and then back in Moscow again—it was a rare person who did not ask, "And why are *you* so interested in all this?" People always asked the question in a polite, puzzled tone: the voice of someone who has recovered from, say, a rare lymphoma, who has lost many loved ones to the same illness, and who is surprised to find a stranger collecting information on the disease when it apparently does not run in his own family.

The more I heard the question, the less certain I was of all the answers to it. My interest did not seem completely rational, nor, in one sense, fair. Why, for instance, was I (like almost everyone drawn to this period of history) so much more riveted by the Great Purge than by the huge famine caused by the collectivization of agriculture, which left thousands of villages full of corpses in Russia and the Ukraine? As far as we can tell, the ultimate death tolls of these two events, give or take a few million, were roughly similar. Is the difference in our interest a matter of class snobbery? The Purge victims included generals and marshals, cabinet ministers and famous writers; the famine killed mostly peasants.

I think a better explanation is something else. Even though this famine was man-made, other famines have plagued many countries before and since. The Purge, by contrast, seems almost totally unfamiliar. It went beyond ideology, beyond reason, beyond factional strife, and beyond even the self-interest of the Soviet Union's rulers. So many engineers and managers were killed that the economy slowed down; so many railway workers were killed that trains failed to run; so many Red Army officers were killed that, a few years later, the Soviets almost lost World War II. The Great Purge reached the realm of madness.

In a tantalizing aside, the historian Roy Medvedev, known for his access to inside sources, tells us that training for NKVD cadets included "a very serious study of the Spanish Inquisition." In what spirit? As an example to be emulated? For learning specific techniques of torture? For tips on how to establish a climate of fear?

There was a certain logic to this curriculum, for the Purge did in part resemble the Inquisition. Like the Purge victims, the accused heretics of Spain were almost all found guilty, were tortured into confessing, and sometimes, in a medieval version of the Moscow show trials, had to confirm their confessions while not under torture. In both countries, the killings ranged high into the hierarchy (Central Committee members; bishops and archbishops). And like the Purge, the Inquisition at first greatly strengthened the grip of the absolutist rulers of the day, but in the end undermined their successors.

The Great Purge, however, far surpassed the Inquisition in its death toll. Whispered rumors of terrible labor camps in places like Karaganda, or the still more feared and distant territory of Kolyma, swept through the country. A sense of dread spread ever more widely. Eugenia Ginzburg and her husband burned their books so they wouldn't be found with something that had been banned. Sensing that they would soon be arrested, they developed a special way of ringing the doorbell, which meant, "It's still only me, not them." The long-feared ring or knock at the door sounds like a refrain through every account of those years, as it had for Vladimir Glebov and for so many other people I talked to. In a recent Russian movie, *Defense Counsel Sedov*, a knock comes as a lawyer and his wife are eating dinner. He reassures her, saying, "It's only ten o'clock. *They* come later." Fearing the knock, Maxim Litvinov, the foreign minister, stayed up until two or three each morning, because he couldn't stand the idea of being arrested in his pajamas. At the height of the Purge, Stalin sent some bodyguards to protect his mother; when they knocked on her door, she thought they had come to arrest her, and fainted.

"People didn't sleep, because every night there were arrests," recalled Alexander Milchakov, a Moscow journalist I talked to who was seven years old when the knock came at his door and his father was sent to Kolyma for fifteen years. "These arrests had one unique feature: the sound of boots booming up the staircase. When the NKVD men climbed the stairs, the tenants racked their brains trying to figure out where they were headed. Would they stop one flight higher? Or lower? They didn't take the elevator—there was some kind of order against this, probably for fear of breakdowns."

Whenever I heard such stories, I always wondered: Why did people wait for the knock? Even if armed resistance was impossible, why not hide, flee, go underground? Another question that made me curious was related to this: When Stalin and his prosecutors made their charges of vast and sinister plots, why did so many people seem to believe them? In 1937, when the verdict in a big Moscow show trial of former Party leaders was announced, some two hundred thousand people assembled in Red Square to hear it—in weather of -27° F. That kind of fervor is built on something besides indoctrination alone.

The propaganda was pervasive, however. The conspiracies people were accused of spread even into the pages of novels. As one account explains: "[When] the third volume of Panfyorov's *Bruski*, a well-known novel about collective farms, appeared. . . . Characters described as noble Soviet citizens in the second volume turned out in the third volume to have been secret enemies of the people."

Like the Inquisition's victims, the millions of people seized by the NKVD were forced to name their supposed associates in heresy. A huge army of interrogators, some three thousand in Moscow alone, worked around the clock to torture and threaten them into doing this. Among all the memoirs of former *gulag* prisoners I've read, I've come across only one account of an interrogator who could not bring himself to do the job:

"He was a young lieutenant, obviously fairly new to the work, who sat at a desk . . . while the prisoner had to stand. The endlessly repeated question: 'Who recruited you?' started growing less frequent, and the stupefied prisoner suddenly saw the examining magistrate burst into a flood of tears. He allowed the prisoner to sit, gave him a drink of tea from his own glass, offered him a cigarette, called for a relief, and the interrogation then continued." There is no record of what later happened to this anonymous young lieutenant.

It is one puzzle why so few people questioned all this madness while it was going on. But it is another why millions of people still do not question it today. Colonel Volkov in Karaganda definitely missed the old days. Back in Moscow, I saw tens of thousands of people who felt the same way at a rally of conservatives outside the Kremlin wall on Soviet Armed Forces Day: older men, mostly, with hard faces and chests paved with war ribbons, sometimes carrying signs with Stalin's portrait or with messages attacking the Jews. What were people like who thought this way today? Did they still believe in "enemies of the people," and all the rest of it?

Kira Kornienkova is a leading figure in a vocal neo-Stalinist group called *Yedinstvo* (Unity). An American friend and I went to see her one Sunday morning at ten o'clock. When I called her, I was relieved that Kornienkova had suggested this hour to meet, because it wasn't a mealtime.

When we showed up at her door, however, Kornienkova immediately ushered us to a table set in her living room, and kept disappearing into her kitchen to fetch dishes. The midmorning meal confronting us included two kinds of salad, sliced smoked fish, slices of something that looked like raw bacon, cabbage-filled dumplings, white bread, dark bread, mushrooms Kornienkova had gathered and pickled, bowls of soup with fat floating on top, strange blocks of jelly with bits of meat at the bottom, and mari-

Kira Kornienkova.

nated garlic. To drink there was juice, homemade wine, and homemade Georgian grape vodka, "*Chacha*—made by the people in Stalin's country!" For dessert there were apples, Russian cake, Georgian cake, and cooked acorns in heavy syrup—a favorite dish of Stalin's, she said.

Why this overwhelming mass of food? To kill potential critics with kindness? To show us that there really are no food shortages in Russia, despite such claims in the capitalist press? Or was it the familiar Russian hospitality reflex, cutting across political lines? Whatever the purpose of the meal, we could not eat enough of it to satisfy Kornienkova. Finally her reproachful urgings gave way to talk about Stalin.

Kornienkova is fifty-five and unmarried. She has traveled all over the country, she said, visiting places where Stalin lived or worked. There are many, because he was often exiled under the Tsars. "I traveled all around Siberia. I went up the Yenisei. I went north, where he had been, to Solvychegodsk and Vologda. The house where Stalin lived is still there. I went to Narym. Even now it's hard to get there. From Irkutsk, you drive and drive and drive. How did he manage to escape from there? And in winter! And dressed very badly for the Siberian frosts." She rattled off a long list of apartments, houses, and dachas where her hero had lived, ticking off on her fingers which ones still had the original furnishings.

Kornienkova's travels were religious pilgrimages. "Every year," she said, "I go to Gori [Stalin's birthplace], on the day of his birthday." And, just as Catholic priests study Latin and devout Jews Hebrew, she has learned her prophet's native tongue, Georgian—no easy feat, for the language is unrelated to Russian and has its own alphabet.

Although she may be a specimen of a mind-set from the Russia of fifty years ago, Kornienkova reminded me greatly of people much closer to home: John Birch Society members I had spent time with in Southern California. One similarity was her obsession with Jewish conspiracies. Another was her pursed-lips distress at the rise of pornography. But the most striking was her constant, exultant, voluminous references to all sorts of written sources. Anyone who has read the outpourings of the American paranoid right will recognize this curious, pseudo-academic style of argument. Its practitioners assume that you can prove any point by the massed weight of citations to documents, no matter whether the document is fact or opinion, applicable or irrelevant, a statistic or a wild charge by another raving Bircher. Kornienkova, too, practically spoke in footnotes.

"Recently I published a piece in *Moscow Builder*, Number 29, last year, the summer issue, where I thoroughly analyzed an article," she said. The offending article was by a liberal historian who had claimed that Stalin was guilty of rape. Kornienkova poured out a triumphant torrent of dates, place names, volume and issue numbers of various magazines, and data from the archives of the Central Committee, all supposedly proving that this episode never could have happened.

But her opponent was crafty, she said. "He has lots of 'witnesses.' He does it very slyly. They *all* do it slyly. For example, he refers to what his mother told him, and his mother heard it from somebody else. But his mother and this other woman died long ago, and there's no way to check whether she said this or not. Out of this kind of garbage they build their books!"

The flood of dates and data continued as she went on zestfully debunking other debunkers of Stalin. She spends her evenings in the Lenin Library gathering ammunition for long letters to the editor. The motives of her enemies, said Kornienkova, are simple: "There is that Yuri Afanasiev [a liberal historian]. Have you heard about him? He's Trotsky's and Kamenev's nephew! And his name is not really Afanasiev, but"—she paused for effect, then drew out the word slowly and triumphantly, an indictment—*"Ros-en-feld!"*

(Today, Kornienkova's anti-Semitism is highly typical of Stalin supporters. Curiously, this was not always so. Stalin himself found other groups of scapegoats, and did not turn on Jews, as Jews, until the very end of his life. Only then did he begin stamping out imaginary Zionist plots, order the arrest of Jewish doctors in the Kremlin, and draw up plans to deport Jews to a remote corner of eastern Siberia. This final frenzy helped unleash again the virulent current of anti-Semitism that has long been part of Russian life. During the blood-soaked Purge years themselves, it was not so visible. Because they were disproportionately represented among groups like intellectuals and Old Bolsheviks, many Jews were victims. But they also numbered high among the perpetrators. Jewish senior secret police officers included Genrikh Yagoda, who was chief of the force from 1934 to 1936. One prisoner's memoir even tells of being questioned by some high-ranking officers in Yiddish, a language he and his interrogators had in common.)

Kornienkova, the person seeing all these sinister Jewish conspiracies against the memory of her beloved dictator, was a sprightly woman with a cheerful smile. She cited her chapter and verse not with features distorted by hatred but with the happy look of someone making a good chess move. Her ample, matronly figure was clad in a white blouse with embroidered flowers. And her love of Stalin included no animosity (nor for that matter, the slightest curiosity) toward two visiting Americans. Most unexpected of all, Kornienkova's professional work, which she seemed very proud of, was as a teacher of retarded and disabled children.

The living room where we were talking was filled with houseplants, small landscape paintings, rag dolls and stuffed animals, and a cage for her two parakeets, Yashenka and Mashenka. From time to time she paused to talk affectionately to the birds and to feed them bits of bread through the bars of the cage.

What about the *gulag*?

Again she cited a volume and issue number of a magazine: "It was published two years ago; I can give you the exact *day*. They arrested some good people, no doubt, but not in the numbers now written about. People who have investigated this issue *in depth* believe that about twenty percent of those arrested were innocent." Otherwise, as far as Kornienkova is concerned, "enemies of the people" really were enemies of the people: "People who were alien to the Party penetrated it, and concealed their past. It was exactly the same as during the war, those who collaborated with the Nazis."

Anyway, she said, as her parakeets chirped in the background, there were not that many people imprisoned; she knew hardly anyone who was arrested. Except, of course, for people who were correctly arrested. A neighbor was locked up for many years—but it was for cowardice during the war. A relative was sent off to the camps—but he had made some terrible mistake at his factory that caused the loss of a million and a half rubles. Two of her uncles were arrested—but "they drank a lot." So why, she asked, do such people deserve "a monument as high as the skies?"

What about the sweeping purge of Red Army officers?

In answer, Kornienkova smiled and touched her finger to her throat in the gesture Russians use when they talk about alcoholism. And besides, she said, "They were not executed, you know. The army was *rejuvenated*. There was a war coming, and the older officers were removed and kept in reserve. Nothing happened to them. A younger generation took their places."

After talking for some time, she took us into her bedroom,

which was also part library and part shrine. Along the walls and on the glass doors of bookshelves were photographs and paintings of Stalin: Stalin at Yalta with Roosevelt and Churchill, Stalin in his study, in a lawn chair, at his desk, reading a book, with his wife, with Lenin, with his daughter.

It would be easy to dismiss Kornienkova as a laughable eccentric stuck in the past. But only forty years ago an entire nation shared her devotion to the man newspapers and banners called "Teacher of the Workers of the World," "Devoted Continuer of Lenin's Cause," and "Greatest Genius of All Times and Peoples." "We loved Stalin. The whole country loved him," she said, with some truth, in a pained, sad tone. "Everyone loved him—and now they refuse to have anything to do with him!"

Again I thought of those John Birch Society members in Southern California. In them, too, there was a strain of nostalgia and sorrow. Most had come to California from the rural Midwest, and the menacing Communist conspiracies they imagined around them belied their own culture shock at moving from the familiar certainties of tight-knit Iowa farm towns to the bewildering sprawl of Los Angeles, where nothing was as it had been, be it neighbors, religion, job security, sexual standards, even the color of the sky. Stalin's Russia, in which a near-feudal peasantry made a sudden jump to urban, industrial life, underwent an even greater sea-change.

But if that passage engendered a suspicious readiness to blame all problems on shadowy villains, it augurs badly for the Russia of today, which is also going through a drastic transition. In a bewilderingly short time, all the old certainties are gone—guaranteed employment, guaranteed good news in the press, the infallibility of the Party, superpower status, the very Soviet Union itself. Kornienkova speaks for millions when, putting aside any notion of cause and effect, she recalls the orderly Stalin era as a happy contrast to today, with its rising crime and corruption, beggars in the streets, newly rich businessmen driving BMWs, ever less food, and an economy in shambles. How many people will respond to all these changes, as she has, by recourse to conspiracy theories, old or new? The precedents elsewhere are not good: the crumbling of social order almost always strengthens the appeal of fundamentalists.

It was hard to make Kornienkova stop. She climbed up on a chair and from the top of a cupboard pulled down a photo album and an unsteady armful of envelopes and file folders with more photos. The album held more pictures than there was room for, photos thrust loose between the pages. They spilled out of her arms as she teetered precariously on the chair: Stalin at age nineteen, Stalin with his brother, Stalin's death mask, Stalin as a seminarian, the grave of Stalin's first wife. Finally, out of breath from her exertions, Kornienkova found the one she was looking for: a group photo of a class of ten-year-old schoolboys.

Smiling playfully, she said: "Find him!"

Unlike the Birchers' fantasies, the delusions of people who thought like Kira Kornienkova served as the building blocks for an entire empire. And the imaginary enemies she believed in were feared even at the empire's very pinnacle. A few weeks after meeting Kornienkova, I got a rare chance for an indirect look at that climate of fear at the top. It was an invitation to spend a weekend at one of Stalin's dachas.

Although Stalin carefully cultivated an image as a man of simple habits, he loved country homes, and at one time or another had a dozen or so, some built to order. One of the most beautiful of these dachas has been, until now, inaccessible. A two-hour flight from Moscow, it sits on a mountainside in Stalin's native Georgia, just across the republic's border with Russia. For several decades after Stalin's death, the dacha was used by visiting top officials of "fraternal" Communist Parties. Starting in 1990, as ties with these parties began dissolving, the dacha was leased as a guest house by a Soviet industrial group. Thanks to Russian friends who work with the organization, my wife and son and I were able to visit one weekend when no one else was there.

Nestled into a high cliff above the Black Sea, the building is completely invisible to passengers on the coastal road 700 or 800 feet below. The snow-capped peaks of the Caucasus slope down to the deep blue water here as spectacularly as the mountains meet the ocean at Big Sur. A fast-flowing creek reaches the sea at this spot, giving the place its name, Kholodnaya Rechka, or "Cold Stream." The steep slopes are covered with cypress, eucalyptus and cherry trees, and with a type of long-needled pine found, people told us, nowhere else in the world. The pebble beach below is fringed with palms.

We reached the dacha by a steep, winding road that finally came to a fence, signs saying "Forbidden Zone," and a gatehouse with a round-the-clock guard. Tucked into its niche carved out of the rocky, forested mountainside, the dacha itself was a large, nondescript, green stucco house in need of new paint. But inside, where most of the five bedrooms had an adjoining living room, it was one of the most beautiful buildings I have ever seen.

Each room was quite spacious, and had elaborate parquet walls, floor, and ceiling made of many different kinds of wood— oak and walnut stained to various shades of light and dark, and mottled birch from Karelia, near the Finnish border. The ribs and panels of the different woods were fitted together in intricate designs, with a lightly lacquered surface. The blend was different in each room, and in each, you could sit back and drink in the rich variety of patterns for many minutes at a time, as if you were enjoying paintings in an art museum. Silver samovars sat on tables. The gold-trimmed curtains and Persian rugs were mostly deep red, drawing out the luxuriant golden colors of the wood still further. The next day we woke up surprised and a little uneasy that we had slept so well. Had Stalin, too, lulled by this great cathedral of woodwork? Were we in his bed?

In the morning we explored the grounds. The mountain air was clear and crisp, and through the pine trees we could see the shimmer of the Black Sea far below. A few outbuildings surrounded the dacha itself: one was a dacha built for Stalin's daughter, Svetlana; another contained pantry and kitchen; and one held a pool table—Stalin loved billiards and his entourage knew that he should always win. There was also a small theater with a projection booth: Stalin was a movie nut, with a special passion for Charlie Chaplin and for Tarzan movies, of which he had a complete set. "The films didn't have subtitles," writes Nikita Khrushchev in his memoirs, "so the Minister of Cinematography, I. I. Bolshakov, would translate them out loud. . . . Actually he didn't know any of these languages. He had been told the plot in advance. He would take pains to memorize it and then would 'translate' the movie."

Maria Nemchenko, a cheerful, friendly woman with a pink bandana, dyed red hair, and a mouthful of gold teeth, waited on table here in Stalin's day. Now sixty-five and retired, she lives in the nearby village. But she was happy to come back to the dacha for what appeared to be the first interview of her life. She had never seen a pocket tape recorder before, and laughed with delight at the sound of her own voice coming out of it.

Before I had a chance to ask her which bed was Stalin's, it became clear: they all were. He slept in a different room each night, she explained. He never told anyone where he would sleep, except "a special woman who came from Moscow with him, a maid. She was the only one who made his bed." Nemchenko walked us through the different bedrooms. Of one she said, in a hushed, affectionate voice: "This was his favorite." It was an odd but perhaps revealing choice. The bedroom was on the side of the house against the cliff; the windows on one side looked out at sheer rock a few feet away. Most of the other bedrooms had beautiful views of the mountains and sea. This one had, literally, its back to the wall.

"Of course the dacha was built by convicts," Nemchenko said. I wondered who they were. Perhaps they were only the unskilled labor, and the master craftsmen were free. Or had somebody gone out and carefully arrested the country's best woodworkers?

Stalin never went downhill to the beach, Nemchenko went on, and when he was in residence he brought with him seven hundred guards. Posted outside on the forested hillside, they surrounded the house in three concentric circles. All the rooms were electrically monitored. "Every door had a wire. The moment you opened the door, the switchboard knew that someone had come into the room." When she carried food from the kitchen across the court-yard to the dining room, an armed guard marched on each side of her. "When we cleaned, they watched us. If I walked, I had to walk without looking around."

Stalin's paranoia extended also to food. He would only eat meat if it were freshly killed. "It was a torture, because we had to do it. Sheep, chicken—everything was alive, and then we slaugh-tered it. A special laboratory traveled with him everywhere, with a staff of technicians. They had rats and mice. They would take a sample and test it, and only after that could we serve the food. And that was three times a day."

Nemchenko was still a believer—a feat she managed by split-ting her ambivalence, as psychologists say, between Stalin and se-cret police chief Lavrenti Beria. "I personally think well of Stalin. He did nothing bad to me. It was Beria who did the bad things. That scum, I hated him!" Beria, a notorious lecher, was always making passes at any young woman in sight, making life miserable for Nemchenko and the other waitresses and maids. "The guards were young guys, they protected me. When they saw him ap-proaching, they would call, 'Masha, hide!' "

At the dacha's dining table, now covered with a white cloth, Nemchenko showed us where Stalin always sat, in the center, sur-rounded by Beria, Foreign Minister Vyacheslav Molotov, and oth-ers. "The airplanes came and went three times a day. Almost the entire Politburo was here."

Stalin spent a great deal of time at his various dachas; visitors describe an atmosphere of forced joviality and drinking, during which the host repeated the same old stories and jokes and the guests were very, very certain to laugh long and hard. One visitor recalls a long evening of drinking games in the 1930s that ended with Marshal Tukhachevsky of the Red Army showing his strength by picking up the diminutive Stalin and holding him high overhead. Such an expression of fear, anger, and humiliation came over Stalin's face that the others in the room all sensed Tukhachevsky was not long for this world. Soon after, he was put on trial and shot.

The evenings that Stalin spent here at Kholodnaya Rechka may have been more restrained, for the dacha was only finished in 1949, four years before his death. That was the period of a sweeping purge in Leningrad and of other new waves of arrests that made many fear a second Great Purge was in the offing. Beria and Molotov, who both dined at this table, were fierce rivals. Beria put Molotov's wife in a labor camp the very year this dacha opened. Soon afterward, he arrested the wife of Alexander Poskrebyshev, Stalin's long-time personal secretary. Reportedly, when Poskrebyshev begged Stalin for her life on his knees, Stalin replied that he could do nothing; it was all in the hands of the police. Stalin also jailed Nikita Khrushchev's daughter-in-law, and two sons of long-time Politburo member Anastas Mikoyan. Every one of these men was in the group that ate around this long table. Did they smile and joke with each other as they passed the *shashlik* or drank toasts? It might have been fatal not to.

"Stalin loved to pour everything himself," Nemchenko said, almost reverently. "I would put the tray in front of him. He would take a cup, pour the tea, and give the tea to everybody."

Besides Khrushchev, few people have left first-hand descriptions of life in Stalin's inner circle. And because he dispatched so many around him to the execution cellars, fewer still survive of those who once sat at Stalin's dinner table. Although what interested me most were the problems of coping with Stalinism's legacy today, I still could not resist the chance to meet one person who remembered what it was like to work with the dictator himself.

He is a white-haired man of seventy-five. One wall of his Moscow apartment is covered with photographs of himself and Stalin together with other world leaders. Over tea and cookies there one morning, Valentin Berezhkov shared his memories of Stalin—in fluent English, for Berezhkov was, for almost all of World War II, Stalin's English-language interpreter.

After studying German and English in school, Berezhkov began working during the 1930s as a guide for foreign tourists. Then he got a job in the Foreign Ministry. Ravaged by the Great Purge, its ghostly offices were filled with empty desks. Since contacts with foreigners were always grounds for suspicion and arrest, the ministry was particularly short of foreign-language speakers. Someone who managed to avoid being shot could advance rapidly. In late 1940, the young Berezhkov translated at secret talks in Germany between Molotov and Hitler. At the outbreak of war, he was working at the Soviet Embassy in Berlin.

The ambassador and chargé d'affaires there, who had known of Molotov's secret talks, were both ordered shot by Stalin. But Berezhkov escaped this fate. Today, he has the wary eyes of a survivor who has skillfully navigated a lucky course through changing political currents, along the way perhaps making more compromises than he wants to talk about or thinks it is within the comprehension of an outsider to understand.

Berezhkov began translating for Stalin in the dark September of 1941, when the invading German armies were fast approaching Moscow. The first time he saw Stalin was when the dictator held a reception in the Kremlin for worried emissaries of the Allies: Lord Beaverbrook from England and Averell Harriman from the United States. "The door opened and Stalin came in. I had quite a shock, you know, because he looked not as I had imagined him, but quite different. He was under terrible stress, quite thin. His clothes were just hanging on him. He had a gray face, marked with small-pox, which was always painted over in photographs. Even in movies the light was somehow arranged so that you couldn't notice it. I couldn't imagine that this small man with one arm shorter than the other and such a gray face, that this was Stalin." Berezhkov interpreted for Stalin for the next three years, and was at his side at the 1943 Teheran conference with Allied leaders.

Were signs of the dictator's paranoia visible to those who worked with him daily? Absolutely, said Berezhkov. "Sometimes he would say, *Why don't you look into my eyes?* Many people had this experience with him. Next time, after this question, I would *always* look into his eyes."

Stalin was "never rude, would never say something offending

or humiliating." But in a very subtle way the dictator always played his aides off against each other. Berezhkov and another interpreter, Pavlov, shared an office and translated any conversations Stalin had with English speakers. Usually Berezhkov translated at talks with Americans, and Pavlov with the British. "But sometimes there would be an English person, like [Anthony] Eden. And all of a sudden I would get a call saying Berezhkov should come and translate. Then Pavlov was always very upset. And then some American would come, and Pavlov would be asked to translate. And then *I* would be upset and would think, Maybe he is dissatisfied, maybe I said something. . . ."

Not everything Berezhkov translated was diplomatic negotiations. "Sometimes they would bring me recordings from the hotel where people from abroad would stay. . . .

"Stalin had a hidden telephone connected with all the Politburo members, so whenever any of them talked to anybody, he could listen. He had it hidden in his desk. He could take it out and listen to anyone. Everybody was afraid.

"Molotov was like his shadow, always present. Sometimes he would say to Molotov [about an aide], 'You know, this man knows too much.' It was immediately a sign that this man was in trouble." Within a few weeks, the man named would disappear. Berezhkov rattled off the names of three or four officials he knew who were shot or sent to Siberia. Among them was the Foreign Ministry's head of protocol, under suspicion for "connections with foreigners"—although that was, of course, his job.

"Sometimes Stalin would say, if something happened, 'You must find the man who is guilty!' A telegram once was not delivered to Roosevelt immediately. Several days passed. Molotov told me I should investigate. I asked the man who was responsible for these coded telegrams. He checked: the Americans were responsible, there was some trouble on their side. So I came to Molotov and said, 'Nobody is guilty.' But he told me, 'How can you say nobody is guilty? Stalin said that you have to find the guilty person and punish him! You are a rotten intellectual if you say that nobody is guilty.' So he asked Vyshinsky to come immediately and said, 'This is the situation and this man is a rotten intellectual, he says nobody is guilty!' "

Andrei Vyshinsky, who never failed to find a culprit when one was needed, took over the case. "And this man, who was head of the coding department of the Foreign Office, he disappeared. I think he was shot."

Was Berezhkov a true believer when he worked for Stalin?

"Of course. We were very much indoctrinated." Today, he said, he feels differently. Yet when I asked him what went wrong, what produced Stalin in the first place, his explanation was personal, not systemic. And it implied that the problem was not that the system produced an all-powerful leader, but only that it produced the wrong one.

"I think those people who considered themselves leaders of the Revolution, equals to Lenin, they were very naive. They underestimated Stalin. They never thought he would become dangerous. I think the important thing was rivalry among them. Lenin did not indicate his heir. He should have understood there would be a struggle."

The Great Purge took place in the late 1930s; after the German invasion of 1941, you would think, Stalin would have had no time for anything besides the Soviet Union's life-and-death struggle with the Nazis. But even when the heartland of European Russia was overrun by the Germans, the country's survival in doubt, and tens of millions of Soviets victims of war, the mania for finding spies and traitors still raged.

"All of a sudden in '44, Molotov asked me, 'What were you doing in the Polish consulate in Kiev in '34?' All connections with foreigners were always suspicious. I said that I was working for Intourist at that time; I would get transit visas for tourists. Molotov said, 'Beria wrote to Stalin that you had some contact with the Polish consulate.'

"This was the first signal. . . ."

Soon Beria raised another charge against Berezhkov: Berezhkov's mother and father had lived in Kiev, which was occupied by the Germans during the war; when the Germans were pushed out, his parents were no longer there. Did this mean, Beria hinted, that they had retreated with the Germans, to be held as hostages, to force their son to become a spy? Or had they been Nazi agents all along? (Years later, Berezhkov found out they had fled to Western Europe.)

Beria's attack on Berezhkov was a maneuver within another, more important attack—on Berezhkov's patron Molotov, the secret police chief's rival. Beria was trying to convince Stalin that the British had recruited Molotov as a spy. This had happened, Beria suggested, when Molotov went to England during the war to meet Churchill and British Foreign Secretary Anthony Eden. "Molotov was accused of being recruited by Eden in England in 1942, when he was going by night train from Glasgow to London. There was a moment when Eden came to [Molotov's] compartment with his interpreter. There was a rule at that time that nobody, even a Politburo member, should ever have a meeting with a foreigner without a witness. This was the only time Molotov did. Maybe Molotov was thinking, 'My own interpreter is sleeping, why should I wake him up?'"

While Beria filled Stalin's ear with charges against both Molotov and Berezhkov, Molotov—a veteran political infighter—moved to protect himself by getting rid of Berezhkov. He told him: "You cannot remain. Turn over your papers and documents. And we'll be thinking what the decision about you will be."

"It was such a blow!" Berezhkov said. "I felt this was maybe the end of everything. The second feeling was, How could I be left out? I was so *necessary*. How could it be that I was not needed at all? The humiliation!

"When I went out the Spassky Gate [of the Kremlin], the man on duty, who knew me of course, looked for a very long time at my [pass] and said, 'I have an order to take it away from you.' I went out, and he shut the door of the Spassky Gate behind me. I went walking around until morning. It was a terrible moment. For two weeks I was waiting every night for them to take me. It made such a good case: a Polish spy, a man whose parents probably cooperated with the Germans, and he was probably a German spy too, because he had been in Germany before the war. . . ."

Although a ruthless man who had signed many a list of execution victims, Molotov saved the life of his young protégé. A few weeks later, Berezhkov got a call: Molotov had arranged a job for him at an obscure magazine. From then until Stalin's death, Berezhkov kept a very low profile, editing translations and writing, at Molotov's suggestion, only under a pen name. Molotov told him: "If you want to live, you must become invisible."

In 1953, Stalin suffered a stroke and died—after lingering helpless and alone for an entire day because his aides, guards, and doctors were all too terrified to enter his bedroom uninvited when he didn't emerge from it one morning. Even though forced by the dictator's paranoia to be "invisible," Berezhkov, like millions of other Soviets, wept at the news of Stalin's death.

Despite an ominous demotion in Stalin's last months, Berezhkov's protector, Molotov, also survived. Molotov was soon back in the saddle as a Politburo member and minister of foreign affairs. Then, for the first time in a decade, on April 18, 1954—Berezhkov remembers the date exactly—Molotov summoned Berezhkov to see him and invited him to rejoin the Foreign Ministry. "He welcomed me as if he had seen me yesterday."

At this point in our conversation—and only at this point, not when he was talking about friends and family who had been arrested or the terrors of the war or the disappearance of his parents or his fears for his own life at the hands of Beria—Berezhkov lost his composure and was overcome by emotion. For a few moments he could not speak.

Why, reviewing a life that had seen so much, was it this one memory that stirred tears? Few emotions are stronger than the desire to belong. And particularly to be accepted back into an elite, closely knit group from which you feel you were unfairly expelled. Stalin and his lieutenants knew this better than anyone, and made good use of it. If there is to be a *them,* a class of enemies, there must be an *us.* It was that desperate desire to get back into the Party at any cost that made so many Purge victims confess to impossible crimes. Berezhkov never had to confess to anything to get readmitted to the fold; but seeing how much the readmission meant to him made it clearer to me why so many thousands of people did confess. Facing the certainty of his own execution, Rubashov, the imprisoned hero of Koestler's *Darkness at Noon*, has looked deep into his own conscience and into the morass of destruction that the Revolution has become. Yet he, too, is uncontrollably stirred when his interrogator, having extracted the necessary confession, at last calls his prisoner "Comrade." From this desire for human solidarity was constructed the machine that destroyed it.

8

"The
Stalin in Us"

It was curious to find Valentin Berezhkov blaming Stalinism on Lenin's failure to name the right "heir." And it seemed particularly striking to hear this apparent belief in strongman government from a sophisticated person who had not only lived in the West, but who himself had come so close to being extinguished by the strongman. Ever since I had visited Karaganda, the lens through which I had previously been tending to see the Stalin period—heroes and villains, noble victims and evil executioners—had seemed much too simple. Drawing those dividing lines is not so easy. Of all the stages in coming to terms with oppression, the hardest is realizing how much we internalize our oppressors. A man treated cruelly as a child is far more likely to abuse his own children than to become a crusader for children's rights. Writ large, the same thing happens: victims begin to resemble executioners. "We should not be afraid of our spiritual kinship with Stalin," writes a Russian historian, Mikhail Gefter. "We must see and realize it in order to eliminate all that is of Stalin in us."

By chance I soon met someone in Moscow who had been thinking about just this point. A major landmark of *glasnost* had been the first big public exhibition about the *gulag,* called "The Week of Conscience." In 1988, it had filled a large exhibit hall with panel discussions, film showings, and displays of letters, documents, photographs, and labor camp tools. I called up one of the people who had organized the Week of Conscience, a man named Alexander Vainshtein, and asked if he still had any of the items that had been on display. They were all in storage, he said, "but I've got some ideas about it all that I'd be glad to share with you." There was something thoughtful in his voice, and I made a date to see him.

Vainshtein is a balding, middle-aged man with a brisk, businesslike air, a journalist turned entrepreneur. He publishes books, has business ties in Switzerland and Israel, and promotes sporting events. His comfortable, wood-paneled office was in a ring of rooms built into the sides of Moscow's Olympic Stadium. He was obviously navigating very well in Russia's uneasy shift to a private economy. Vainshtein's work arranging the Week of Conscience, however, had not been a business proposition but a labor of love.

"One's attitude toward Stalin," said Vainshtein, "is a sort of litmus test of one's attitude toward life in general." Setting up the exhibit, as he looked back on it now, some two and a half years later, seemed "a kind of euphoria. We were like kids who had been locked in the basement, and all of a sudden were allowed to play on the lawn, to run around, to shout, to be happy. There was green grass and everything was great."

But his experience ended on a more gloomy note. "I realized for the first time that it would not be so simple to change things in this country. We had a big book where the visitors wrote down their reactions. It had a lot of aggressive entries, like: 'All the current political leaders should be shot!' And these people think they're democrats! It's the same totalitarianism all over again."

Vainshtein was even more depressed by what he saw after the exhibit shut its doors. Word of the Week of Conscience had spread and people who wanted to see it continued to arrive in Moscow from other cities. Most were *gulag* veterans or their families. But the week for which Memorial had rented the hall was over. "There were all these old babushkas, the ticket-takers—in this country that's a low-paying job and usually retired old women do it—and during the week they'd worked terribly hard, letting in thousands of people every day. They'd been working from eight in the morning until nine at night. They all had realized that this event was very necessary, very important, and they had ignored their own needs. But when they started to explain that the exhibition was over, these people who'd arrived late began to attack them: 'You Stalinists! We'll call the police!' This was the greatest shock for me. So much malice has accumulated in people. I don't know how many generations have to pass to change all that. . . .'"

After the initial ecstasy of reveling on that green lawn of freedom, to use Vainshtein's metaphor, many Russians since about 1989 or 1990 have realized that giving up those old, authoritarian habits will be far slower and harder than anyone imagined. I must have met half a dozen people who told me the same parable: Why did Moses spend forty years taking the Jews through the desert, when he could have made the trip in a week? Because he was waiting for all those to die who remembered the habits of life under slavery. Alexander Herzen, the great nineteenth-century Russian thinker, understood how hard such a transition would be: "To dismantle the Bastille stone by stone will not of itself make free men."

Writing recently, a philologist, Ludmilla Saraskina, pointed out one symptom of this problem. She was discussing the language used in the newspaper of one of the first of the independent political groups to spring up under *glasnost*. The organization claimed Christian Democratic politics, but "the newspaper is unrivaled in its lavish use of words like 'eliminate,' 'unmask,' 'liquidate,' 'dispossess,' 'put an end to,' 'bring to court,' 'political prostitutes.' . . . Not to properly argue with the opponent, but to put a derogatory label on him, to refuse to see even the possibility of good intentions and honesty in the opponent, to see evil-mindedness in him—that was one of the most characteristic traits of Lenin's style of polemics."

Sergo Mikoyan, a historian whose father was a member of Stalin's Politburo, writes of an experience with the same lesson: "A year or two ago at a public discussion I received a note with the following text: 'You and the son of Khrushchev should not write articles or speak at any gathering; instead, both of you should be imprisoned for the crimes of your parents.' I read the note aloud . . .

and asked the audience if they perceived any difference between that thought and Stalin's policy of killing and/or imprisoning the wives and families of 'enemies of the people'."

In one of his short stories, Varlam Shalamov, the writer who survived Kolyma, provides a haunting symbol of the persistence of old habits. A prisoner saves himself from being worked to death in the gold mines by injuring his hand, so it has to be amputated. In the hospital, he can still feel the sensations of the hand that is no longer there. And what he feels are his fingers bent around the handle of a pick or shovel, the tokens of his enslavement.

What, then, is the shape of that hand's grip—the habits of thinking that dominated Russia so disastrously a half century ago, whose traces still linger on in the Russia of today, the "Stalin in us"? As I began to put this question to people, and tried to follow discussion on the subject in the mass media, several themes emerged.

The most obvious one, of course, is the old Russian tradition of absolute power at the top and passive obedience at the bottom. Until Peter the Great made a law against it, citizens of the capital would lie down in the street when the Tsar's carriage passed. You could compile an encyclopedia of such features of autocracy, but the most important one was that until the middle of the last century, nearly half the Russian population were serfs. For almost all the Soviet period, historians treated serfdom merely as one more evil of the old order. Today, however, writers are much more likely to talk about it as the source of a deep passivity that allowed dictatorship to flourish long after serfdom itself had ended.

A recent Soviet short film, *Adonis XIV,* presents a stark image of the connection between that ancient passivity and the horrendous death toll of more recent years. Adonis is a goat. With bells fastened to its horns, it leads a herd of animals to the slaughterhouse. Then a worker removes the bells, and Adonis is slaughtered like the others—tied by its legs to an overhead moving belt, its throat swiftly cut. A slaughterhouse worker mounts its horns on a wall alongside other pairs of horns labeled "Adonis I," "Adonis II," and so forth. There are no voices in the film, only the sound of the animals and the tinkling bells.

Another important ingredient of totalitarian culture in Russia is the country's long love affair with creeds that promise a millenarian deliverance from all suffering. You can easily trace this strand from the schismatics of the medieval Russian Orthodox Church through some of Dostoevsky's characters to the rigid arrogance of the early Bolsheviks. Bertrand Russell wrote of his visit to Russia right after the Revolution: "The governing classes had a self-confidence quite as great as that produced by Eton and Oxford. They believed that their formulae would solve all difficulties."

The problem comes, of course, when the formulae don't deliver the goods. Nothing is more dangerous than the combination of enormously raised expectations and a big drop in the standard of living. The brilliant Polish writer Ryszard Kapuściński noted just such a moment after the overthrow of the Shah of Iran:

Demonstrations began, the crowds shook their fists. But against whom this time? ... Millions of people were out of work, the peasants were still living in miserable mud huts. ... You don't have anything to eat? You have nowhere to live? We will show you who is to blame. It's that counterrevolutionary. Destroy him, and you can start living like a human being. But what sort of counterrevolutionary is he—weren't we fighting together only yesterday against the Shah? That was yesterday, and today he's your enemy.

This same pattern unfolded in Russia. "The Bolsheviks promised this bright future where everything would be so wonderful for everyone," an editor at *Moscow News* told me over dinner one evening. "The promise was so strong, so certain, that when it didn't come true, then *someone must be to blame.*" Ordinary Russians were all the more ready to believe in stealthy conspiracies because under both Tsars and Communists, all important political decision making went on behind closed doors.

As Stalin tried to force-march his enormous country into the industrial age in a few short years, it was severely hobbled by shortages of skilled workers, trained engineers, foreign exchange, and much more. Impossibly high quotas and overcentralized planning made for canals too small for the right ships and parts too big for the right machines. Consumers ran short of everything from bread to soap, and hastily built factories were crippled by accidents. All this just increased the need for scapegoats. The amazing thing is how conscious and explicit the search was. "Assembly lines do not stop by themselves, machines do not break down by themselves, boilers do not burst by themselves," intoned *Pravda* ominously in 1937. "Somebody's hand is behind every such action. Is it the hand of an enemy? This is the first question we should ask."

The urge to find somebody else to blame for one's troubles is not unique to Russia, of course. Hitler blamed the Jews, McCarthy blamed the Communists, Amazon Indians blame malevolent shamans of other tribes; the list of scapegoat-seekers, great and small, goes on without end. But Stalinist scapegoating had particularly deep roots to draw on. In Russia, it has always been a short step from that millenarian promise to persecuting people accused of standing in the way of its fulfillment. Major monasteries often had their own prisons, used for locking up religious heretics. (While a seminary student, Stalin himself spent time in such a cell, for reading forbidden books.) One monastery, complete with jail, on the Solovetskiye Islands in the White Sea, became the site of one of the Bolsheviks' first major prison camps. If you want to understand Russia, a young Memorial scholar studying the islands' long history told me, that's the kind of place where you have to begin.

The scapegoat theme is explored brilliantly in the film *Chuchelo* (*Scarecrow*), one of the finest Soviet movies of the 1980s. The plot involves a class of schoolchildren, eleven or twelve years old, in a small riverside town. A kind and decent girl named Lena is a newcomer to town. The other children tease her for her angular, awkward looks. One day when their teacher is unexpectedly away, the entire class plays hooky and goes to the movies. Someone tells the teacher, and the class is punished. Vengeful and bullying, the children search for the tattletale. They finally frighten one student into admitting to the crime. Lena knows that this boy is innocent. She can't bear to see someone falsely accused. To spare him, she herself claims that she told the teacher—expecting that this will shame the real culprit into coming forward. It doesn't. The class cruelly persecutes Lena, burning her in effigy, and taunting and kicking her in full view of a group of tourists strolling through the town, who pay no attention. While this allegory of show trials, forced confessions, and "enemies of the people" unfolds, a military band booms out cheerful marches in a local park.

"The principle of 'look for those who are guilty' is still very much alive," said Nikita Okhotin, the head of historical research for Memorial, "in all sorts of ways. Somebody will go into a shop and yell at the sales clerk, 'No butter again! You've stolen it, you son of a bitch!' While he doesn't remember that *he,* for the last ten years, has just been drinking tea in his office, doing nothing useful. That's why it's so tempting for him to find an enemy: 'There he is! Right next door! He's stolen my butter!' "

There are other versions of this, Okhotin said, in which a group is accused, instead of one person. "We intellectuals have always worked hard, but those peasants are all alcoholics, twiddling their thumbs. Everybody else is guilty—except my group."

Finally, one thing that distinguished the culture of Soviet totalitarianism from that of many others was the insistent pressure for everyone to join in the scapegoating. It was never enough for the NKVD merely to brand men and women "enemies of the people" and shoot them. The director of any school, factory, or office had to countersign the papers ordering an employee's arrest. And when the victims were prominent, millions of people joined in condemning them by signing group telegrams or open letters, or by voting for unanimous resolutions at compulsory meetings at their workplaces.

When Lev Kamenev and Grigory Zinoviev were sentenced to death in 1936, a voice in the chorus belonged to their one-time comrade-in-arms, Nikolai Bukharin, who wrote, "I am terribly glad . . . that the dogs have been shot." When Bukharin himself went on trial two years later, people who were pressured to sign letters demanding his death and the death of his co-defendants included most of the country's leading writers, several of whom would later be persecuted or shot themselves: Boris Pasternak and Mikhail Sholokhov ("We demand the spies' execution! We shall not allow the enemies of the Soviet Union to live!"); Yuri Olesha and Isaac Babel ("No mercy to the Trotskyite degenerates, the murderous accomplices of Fascism!"); Vasily Grossman and Mikhail Zoshchenko; and many more.

In a society that demanded group condemnations, many rushed to condemn before even being asked. Nadezhda Mandelstam, the most acute of all observers of such things, noted a telling example of this behavior by a bureaucrat's wife soon after the Revolution. The well-known poet Nikolai Gumilev had been arrested and accused of being a Tsarist agent. "One day Legran . . . told us about the execution of Gumilev. . . . His wife, to all appearances a pleasant and friendly woman, butted in to say that she had never liked Gumilev, describing him as standoffish, rude, and impossible to make out. Legran's wife was thus something of a pioneer: in those early years, people had not yet learned to disown the victims so quickly, saying—in all sincerity—what bad characters they had been, and how wrong-headed. Later everyone behaved like this. . . ."

The practice of demanding group condemnations continued, in much milder form, long after Stalin's death. Andrei Sakharov was the greatest dissident of the post-Stalin years; in the 1970s and 1980s, many of the country's prominent scientists signed statements condemning him. By this point, the penalty for not signing was no longer the *gulag* or "nine grams of lead"—the executioner's bullet—but the loss of a promotion, of the use of a summer dacha, of a research fellowship, or, especially, of the much-treasured passport for foreign travel. But the emotional effect was the same: many a broken friendship, and a corrupting sense of complicity that spread through the whole society. When Sakharov died in 1989, an anonymous wreath at his funeral bore the legend: "Andrei Dmitrievich, forgive us."

Until a few years ago, most Russians willing to talk about all this would stress that those millions of people who demanded death for the mad dogs did so only because they had no choice. To any who might refuse, the icy hand of Kolyma beckoned. And that is true. But the tenor of this discussion is changing, acknowledging more of the "Stalin in us." As the novelist Tatiana Tolstaya wrote recently:

> Without popular support Stalin and his cannibals wouldn't have lasted for long. The executioner's genius expressed itself in his ability to feel and direct the evil forces slumbering in the people; he deftly manipulated the choice of courses, knew who should be the hors d'oeuvres, who the main course, and who should be left for dessert; he knew what honorific toasts to pronounce and what inebriating ideological cocktails to offer. . . .

Whether you talk to people who lived through the Nazi period in Austria or the time of the generals in Argentina or communism in Czechoslovakia, today almost everybody describes themselves as victims. The role is always a comforting one. But most absolutist regimes that long remain in power can do so only with the collaboration of their subjects. In the heart of every serf lurks an identification with the master. The Marquis de Custine, wisest of all travelers to Russia, realized this more than 150 years ago, when he wrote that "sovereigns and subjects become intoxicated together at the cup of tyranny. . . . Tyranny is the handiwork of nations, not the masterpiece of a single man."

How much of Stalin is in *us*, outside Russia, as well? How easily might any of us become denouncer or dictator if our society allowed it? I found myself thinking about this after spending an evening with Alexander Vologodsky.

Vologodsky is a bearded, very soft-spoken Moscow physicist who works with DNA. The great passion of his spare time, however, lies in a remote tract of Siberian marshland, north of the Arctic Circle. Some twenty years ago, as a student, Vologodsky had a summer job on a tugboat that shepherded rafts of logs down the Yenisei River. The Yenisei is one of several long, broad rivers that cut through Siberia from south to north. As his tugboat was passing one day through monotonous stretches of uninhabited swamp, Vologodsky looked ashore and noticed a deserted settlement: "It was almost a wooden town; there were big buildings and watchtowers."

Boatmen told him that it was the terminus of an abandoned railway. Vologodsky pulled out an atlas to show me where the Yenisei reaches the Arctic, and then, some distance to the west, where the next major river, the Ob, flows into the same ocean. Except for a few dots marking tiny settlements, the map between the mouths of the two rivers was blank. The plan had been to build a railroad, more than 800 miles long, across the empty territory between these two rivers.

After that summer, Vologodsky quietly began doing research. At a cabinet meeting in 1947, he discovered, Stalin said that he wanted this railroad. Construction began the next year. All the laborers were prisoners. In 1953, a few months after Stalin died, all work on the railroad abruptly stopped.

Vologodsky found himself unable to forget those abandoned watchtowers. For many years it was impossible to investigate such things openly, but since *glasnost,* he has returned to the ghost railroad every summer. "Each time I thought, I won't go back," he said in his very quiet voice, "but I did. And in some four weeks, I'll be going there again." This new trip will be his first in spring, to take photographs without the thick summer foliage. Why, I asked, was he so fascinated by these remains?

For two reasons, Vologodsky explained. The first was that because of the isolated location, the railroad and the prison camps that housed its builders "are preserved—almost like a nature reserve or a national park." Almost no one lived in the region before, and almost no one has lived there since.

Why, then, put a railway there? That's the second reason for Vologodsky's fascination. As far as he can figure out, there was no

logical reason to build this long and expensive railroad—
particularly in the famine-ridden, ravaged, exhausted USSR of
1948, which had just lost one out of every six citizens in World
War II, and where millions of the survivors were still living in cellars and ruins. The Soviet Union is famous for grand public works
projects that turn out not to work; but this Arctic railway, said
Vologodsky, was "the acme of the absurd."

In the frantic haste to satisfy Stalin's orders, Vologodsky said,
when construction began in 1948, "they were laying the tracks at
the same time as they were surveying." The terrain was a builder's
nightmare: below ground was rock-hard permafrost; on top of this
lay six feet of snow in winter, and, in the summer, vast bogs that
swallowed up ties, tracks, and equipment. Although the work force
of prisoners reached as high as one hundred thousand, in five years
they succeeded in laying tracks over little more than half the route.

Because Stalin was so eager to get the railway built, "the death
rate was not high. It was much better supplied than other *gulag*
sites. Of course conditions were severe, people died, and people had
scurvy, but so far as I know, people did not die of starvation there."
Even in summer, though, life building the railway was no picnic.
"There are lots of mosquitoes. People tried to escape from these
camps from time to time. Usually, they were caught and executed.
Sometimes they would merely undress them and make them stand
naked under a watchtower. They'd die within several hours from
the mosquito bites."

Today, thinking of this waste of resources and human life, it
seems easy to condemn the folly of this railroad. But listening to
Vologodsky talk, it occurred to me that in other parts of the world,
when such projects reach their aim, we often honor them as great
feats of engineering or symbols of national grandeur. The Pyramids; the First Transcontinental Railroad; the Panama Canal. Between these efforts and something like Stalin's Arctic railway,
where do you draw the moral dividing line? It is not always easy.
How many slaves died to build the Pyramids? Archeologists only
recently discovered hundreds of hastily buried skeletons in mass
graves nearby.

For the modern slave laborers who built the Arctic railroad, there were more than eighty prison camps all told, strung out at intervals along the track. Vologodsky pulled out a big flat cardboard box of black-and-white photographs he has taken, and spread them out. Not only were the prison camp buildings intact, but even the signs and posters in them still displayed their slogans to rooms full of wooden bunks empty for forty years:

LOVE BOOKS, THE SOURCE OF ALL KNOWLEDGE.

DON'T MAKE NOISE—ALLOW YOUR COMRADES TO REST.

RAILWAY WORKERS! KEEP THE RAIL BED IN GOOD CONDITION. DO NOT ALLOW TRAINS TO DERAIL.

There was an eeriness to these pictures. A camp watchtower, surrounded on all sides by fast-growing birch trees now higher than the tower itself. A rail bed turned into a swamp, that had rotted the ties and then receded, leaving iron rails suspended and bowing, a foot or two above the water. Other tracks with trees growing between them. A semaphore signal hanging at a crazy angle. An entire train, heaved off the track and onto its side by the succession of frost and thaw.

For a long time afterward, I was haunted by these photographs. The more I thought about them, the more they were replaced by another picture: the image of Stalin, at that meeting of his ministers in 1947, pointing at a map and saying, "I want a railway—*there*." Yet when I thought of the finger resting on the map, it did not feel ... foreign. I loved maps as a child, and often planned imaginary railways and canals, to close circles that nature or human builders had, in their small-mindedness, left open. Maps showed a rail line in Alaska, for instance, that was annoyingly separate from the larger network in the rest of the United States and Canada. Why not connect them? And why not a canal between the Rhine and the Rhône?

All this was the impulse, I think, grandly to impose order on a disorderly world. It was precisely this aspect of early Soviet communism—the Five-Year Plans and the great new railways and dams, the White Sea Canal and the Moscow-Volga Canal—that appealed to so many of the fellow travelers who visited the country.

It appealed especially to the late Victorian apostles of rational, orderly progress like George Bernard Shaw and the Webbs. To them it seemed even more splendid that these new projects tamed both the earth and the "antisocial" prisoners put to work on them.

The finger on the Kremlin map suggested a deeper level of answers to the question Russians kept asking me, "And why are *you* so interested in all this?" Totalitarian states fascinate us because of the way they reflect buried parts of our fantasy life. The child imagines new canals on the globe; the adult travels to a distant dictatorship and comes home to praise its canal-building. We are attracted to Utopian visions because they suggest how we can remake the world—and because they suggest how *we* can remake the world. The essential wish to create a better, more just society uneasily shares space, in our hearts, with the wish to wield the power for such creation. The ruins of Stalin's Russia are a museum of what happens when a country loses sight of the first, and gives in to the second. What is left is a railroad from nowhere to nowhere.

9

Beyond
Black and
White

Looking back at the events of the Stalin years that Russians are now dealing with gives rise to a sense of awe. Throughout history, human beings have periodically slaughtered people of other countries, races, and religions. But self-inflicted mass murder is far more rare. How could a country do this to itself?

Our awe also comes from the scale. In the late 1930s, according to figures cited today by Russian government officials, more than one out of every eight men, women, and children in the Soviet Union was arrested—the great majority of whom were shot, or died in prison. In Moscow, the city's crematoria worked overtime. "There was this funny ash that fell from the sky every once in a while," one man told me. And beyond the scale of death, our awe is also at the spectacle of an entire society believing in plots that now seem to us paranoid fantasies. What Russians have to come to terms with today is not only a past of horror and suffering, but also of mass delusion.

Stalin's biographer Alex De Jonge tells the story of four Arctic explorers (one an NKVD man keeping an eye on the others) and their dog, who were stranded on an ice floe during the Purge. "The explorers and the dog celebrated Stalin's birthday and other Communist festivals by holding demonstrations on the ice, marching up and down with banners, since none of the quartet would dare suggest that the activity was preposterous. In the meantime, they got news of the purges on the radio and learned that all those who had sent them on their mission, their replacements and the replacements of those replacements were foreign spies. . . . By the time the team was rescued they all were half-mad, trusting no one, not even themselves and giving rise to a joke popular at the time: 'I'm alone on an ice floe, and there must be an enemy somewhere.' "

The pervasive fear was even more intense because people seldom knew exactly what happened to those who disappeared. The few who returned were afraid to talk. A retired doctor described what it was like as she was going to medical school in the late thirties:

"Every day we found out that someone new was missing. We would come to classes, and there would be no Poles. They all disappeared. And Russians, too, of course. The Dean. The Minister of Public Health. Lots of others around me.

"In a student hostel room, another girl had recently been released. She would wake up in the middle of the night crying, 'No, no, no, no!'

"We'd soothe her and ask, 'What's the matter? What have they done to you?'

"She said, 'I signed a statement promising that I'd never talk about it.' "

"There were all these secret, unmarked doors in train stations and other places," another person who lived through the time remembered. "People walking along the street would look the other way if someone was taken through one." It was as if an entire society were acting out the most extreme of persecution manias. Even including the unmarked doors. In Kafka's *The Trial,* Joseph K. opens the door to what he thinks is a storage closet at the bank where he works, only to find two employees being beaten by the burly, mysterious, leather-clad Whipper.

" 'Is there no way of getting these two off their whipping?' K. asked him. 'No,' said the man, smilingly shaking his head."

In Russia there was no such way, either. Ernst Fischer, an Austrian Communist who lived in Moscow during the thirties, writes:

The most absurd statements, the most implausible lies begin to take effect if repeated day in, day out. Arrests and accusations on such a scale *cannot* be the result of pure arbitrariness . . . once an apparatus is in motion, it is bound by its very nature to gather momentum, and its progress cannot easily be controlled. Vigilance! Are you blind? Can't you see the enemy? Anyone may be an enemy, unless you know him inside out. Vigilance became a matter for competition. Haven't you discovered an enemy yet? You mean to say your organization's the only one without an enemy? How strange, how suspect!

The paranoia penetrated as deeply into the ruling Soviet elite as it did into the rest of the society. Andrei Vyshinsky, chief prosecutor at the Great Purge show trials, the man who found spies and traitors everywhere, survived to become Soviet Ambassador to the United Nations after World War II. In his seventies, he died of a heart attack in his study on Park Avenue. When his wife heard the news, she screamed, "They've killed him!"

Grigory Roginsky, one of Vyshinsky's deputies, got swallowed up by the *gulag* himself. But even in camp, he continued to argue with the other prisoners, defended the mass arrests, and sent home appeals for still harsher punishments for enemies of the people. The writer Eugenia Ginzburg met another such imprisoned official in a communal cell, who took her aside and warned her, "Treason—appalling treason . . . has worked its way into every branch of the government and Party. . . ."

When outsiders were caught up in this madness, they were baffled. In a rare episode just before World War II, the authorities caught a real spy, a Polish agent. The man found himself in a cell in Minsk crowded with Soviet officers and Communist Party members—all accused of being spies. He asked one of the military men, "After all, I am a Polish citizen, a Polish nationalist, an officer and a patriot, in a Soviet prison. That is normal; that is absolutely normal. But why are you, a Soviet patriot and Communist, in a Soviet prison?" No one could explain.

One group of prisoners was accused of plotting to overthrow

Soviet power with Lawrence of Arabia. Another supposed conspiracy was composed of children: some one hundred of them down to the age of ten were arrested in the town of Leninsk-Kuznetsky. "One ten-year-old," writes Robert Conquest, "had broken down after an all-night interrogation and confessed to membership in a Fascist organization from the age of seven." There was no limit to the imagined conspiracies. The patent absurdity of such charges was, in a way, almost a taunt to the public, as if the regime were asserting control of logic itself.

In any society with a recent past of genocide, one of the most important questions is: What are you going to teach schoolchildren about it?

For thirty-five years after Stalin's death, the answer was, essentially, nothing. Nowhere was after-the-fact denial of reality more blatant. Every student across all eleven time zones of the Soviet Union used the same textbooks. Schools teach twentieth-century Russian and Soviet history in the final two years of high school. When the latest pre-*glasnost* edition of the textbook for the first of these courses (up through 1937) came off the presses in 1986, the print run was 3,890,000 copies. The book has exactly ten lines about Stalin. It praises the collectivization of agriculture, ignoring the 7 to 10 million people who starved to death as a result. It does not mention the Great Purge. This ignoring of history is as if the Holocaust did not appear in a German textbook. And, because the Soviets were the main victims of the Purge as well as its perpetrators, it is also as if the Holocaust did not appear in an *Israeli* textbook.

It was this kind of thing that led the reformist historian Yuri Afanasiev to say that the textbook version of Soviet history left children graduating from school feeling that no adult could be believed. How could it be otherwise, he writes, when "we started to call everything by the wrong name. Totalitarianism was called democracy, things that were not begun were proclaimed to be finished, people digging a hole were called people storming the sky."

As the shock waves of *glasnost* spread across the nation, no single event was more dramatic or telling than one that came in May 1988. The Soviet government abruptly announced that, for the entire country, the annual high school exams on the country's twentieth-century history would be canceled. The government newspaper *Izvestia* was blunt about the reason: the standard textbooks were "full of lies."

The educational establishment, however, was no more flexible than the rest of the Soviet bureaucracy. The new flood of debate that filled newspapers and magazines percolated only very slowly into the nation's classrooms. A partially revised textbook that challenged Stalin but not Lenin appeared in 1989. But even that does not mention the Nazi-Soviet Pact, and says the Baltic states "volunteered" to join the USSR—interesting news to Lithuanians, Latvians, and Estonians who watched the Red Army march into their countries in 1940. A few other new textbooks have appeared since, but real change will take a long time, for it means training an entire new generation of teachers.

Meanwhile, what is happening in the average high school history class? I visited several to try to find out. I was particularly looking forward to one class in an elite school for pupils talented in math and physics. It was a small seminar called "Sources and Mechanism of the Stalinist Totalitarian Regime." Such a course would never have been offered a few years earlier. Exactly what I was looking for, I thought: a great chance to see how a reform-minded school was now boldly tackling those thorny issues head-on.

The homework for the class period I sat in on had been to study a booklet of readings. Even though there were less than half a dozen of them, the students sat in a row facing the teacher, who remained behind a desk. One after another, each student, looking stiff and anxious, spoke for ten minutes or so in a rapid monotone, summarizing the main points of an article from the booklet. "Svetlov says . . . Svetlov says . . . Svetlov thinks that there are four reasons for . . ." Then the student stopped, relieved, and another boy or girl carried on: "Borisov says ... Borisov says ... Borisov names three factors leading to ..."

At the end of one boy's spiel, the teacher asked him, perhaps for my benefit, "And what do you think?"

"I don't know," replied the boy, sounding alarmed, as if the teacher had never asked such a question before. Without pursuing the point, the teacher called on the next student. From time to time, the teacher himself spoke for a few minutes. There was no discussion whatever. When the class was over, it occurred to me that the choice of subject matter was irrelevant. There could be no better illustration of the Mechanism of Totalitarianism than the class itself.

How do you begin to change all this? How do you challenge the idea that one man—tsar, dictator, or schoolteacher—is the source of all wisdom? Someone suggested I go and see a man named Igor Dolutsky. He had something to do with writing new high school history textbooks.

Dolutsky corrected me. The problem, he explained, as we talked in his Moscow apartment one sunny morning, is with the very idea of a textbook. Even the most enlightened one is no good, he said, if everybody treats it as simply the new official point of view, which students then recite in class as rigidly as the old. Dolutsky, a former high school history teacher himself, has instead devised a series of books that *have* no point of view. They are collections of documents and questions. He believes that this is the only way he can leap over the heads of tradition-bound teachers and give students the subversive idea that they themselves have the right to make their own judgments on history—or on anything else.

To show what he meant, Dolutsky picked up a small paperback book he had edited. His thin, almost bony face held eyes that glowed with the warmth and animation of a born teacher, someone for whom the greatest of pleasures is forcing others to question their assumptions. He turned to the beginning of the book, a chapter on the early 1920s, when what remained of the democratic im-

Igor Dolutsky.

pulses in the Bolshevik movement were ruthlessly stamped out.
"Here is one proposed [Communist Party] platform, here is another, here is a third one. I ask: Which platform would *you* prefer?
And I don't name the authors. Why? If five years ago I'd written that
one platform was by Lenin, the students all would have chosen it.
Today, if I mentioned that it was by Lenin, they'd all be against it.
That's why I don't say.

"Our education is designed so that the author of a textbook, or
a teacher, imposes his views on a student. I'm a good teacher, the
kids can see that, and they trust me. But another good teacher
might be a member of Pamyat [an anti-Semitic Russian nationalist
group]. And they'd trust him. That's why I say, in effect: 'Listen,
kids; you shouldn't trust me. Investigate for yourselves!'

Some of the book's pages were on greenish paper, some on blue
paper. Was there some design in this color scheme? Not at all,
Dolutsky said; this was the only paper available. He has little bureaucratic clout; this volume and several similar ones have been

published only in minuscule editions, sometimes photo-offset from typewriter type.

The educational establishment still resists anything so innovative, and the authors of the old textbooks are jealous of losing their royalties. Turned down by the main schoolbook publishers, Dolutsky had to bargain with a private "cooperative" to get these books printed. Teachers hungry for something new have pirated and photocopied his books as far away as Siberia and the Arctic. In one sense, this is good; but it deprives Dolutsky of his royalties, and having been forced out of his teaching job in economically difficult times, he needs these.

Dolutsky stressed, however, that the last thing he wants is for his books to be the only ones. "Let there be an abundance of books, so a teacher can choose. Let it not be only Dolutsky or only Kukushkin [principal author of the old, conservative textbook], but another, and another. And if I'm a teacher, I come to a bookstore and choose."

In those blue and green pages on the 1930s, the documentary material Dolutsky has assembled includes everything from speeches by Stalin to poetry by Osip Mandelstam, one of his victims. All through, Dolutsky intersperses questions to the student, using the familiar form, the second-person singular. His chapter about the Purge years ends: "And what do *you* think? What would have been your position, and actions, in the thirties? Maybe it would have been better to have done something earlier? What? When?"

One danger, Dolutsky kept stressing, is today's tendency for both teacher and student to see everything in reversed colors. "One myth is substituted for another: before, the Tsar was bad, now the Tsar is a saint. Well, let's see. In January 1918, the Bolsheviks shot demonstrators who were supporting the Constituent Assembly [the first democratically elected national legislature] and they dissolved the Assembly. They were bastards, weren't they? Yes, they were. Terrorists, murderers! Well, okay"—his face showed the pleasure of pitching a teacherly curve ball—"but what did the Tsar do in 1905, also in January? People were appealing to him, they were petitioning him. He also shot them down! Why are [the Bolsheviks] terrorists and this man a saint?"

Dolutsky grew up in a conventional family; his father was an army officer. But in the 1970s, he was kicked out of Moscow State University for asking too many pointed political questions in a seminar. Drafted into the Red Army, he read *samizdat* literature while on telephone-answering duty at night. Dolutsky's wife is a Latin Americanist, and she borrowed library copies in Spanish of books that were banned in Russian. Dolutsky also collected a large file of materials on "unofficial" history, but then came a KGB crackdown. "In 1982, I burned everything. We thought our apartment was going to be searched."

Yet as a schoolteacher, even long before Gorbachev, Dolutsky felt able to talk to his students freely. "Children would not report me. We had a good relationship. They noticed that I wasn't telling them that Comrade Brezhnev is a leader of genius, and they liked it. They trusted me and I trusted them. I had several students whose parents worked for the KGB. One boy told me, in strict confidence, 'They've started a case against you.' His father had learned that teachers from the school had reported on me."

Talking to Dolutsky was exhilarating. Despite all the centuries of passivity and fear that have discouraged so many Russians from learning to think for themselves, he made you feel, *it can be done*. Even well-intentioned, liberal-minded Russians I met often seemed depressed that their past has turned out so much more complicated than the black-and-white truths they learned in school. But Dolutsky reveled in this, energized by every contradiction and paradox, by every piece of evidence that history is not simple and the answers not easy. His bony face came more alive than ever when he described one thing he had done. The episode had lost him his last job as a schoolteacher, but to him that wasn't the important point.

Dolutsky organized his students to put on a trial of the early Bolsheviks and of Stalin. "We all had roles. There were judges and their assistants, who sat at a podium made of desks. There were lawyers. There were [defense witnesses]: an Old Bolshevik, a Party official, and so on. There were prosecution witnesses: Tsarist generals, a member of the Tsar's family. We distributed the parts." Dolutsky pointed at various spots around an imaginary classroom. "I assigned all the kids to prepare scripts for the characters they were playing. They did everything themselves.

"We heard testimony on the Bolshevik terror, on how the Party was formed; we investigated Lenin's responsibility. The next step was Stalin. A student playing the role of a Tsarist general sided with Stalin. Why? Because Stalin had created an empire! The students brought in magazines and journals dating from those times. Things I'd never ever seen. We invited the parents and other teachers. The audience had the right to interrupt, but they all just sat there; they were scared. The teachers reported on me again. After that, the principal refused to renew my contract.

"But everything went well: the kids weren't paying any attention to the teachers and parents. They were performing. The Old Bolshevik was tearing his shirt. The chairman of the jury was watching to see that everything was done according to the rules."

Most impressive of all was the role Dolutsky had assigned himself in this trial. For it showed that he was not only trying to teach his students something about the facts of history, but also about the rule of law, the principles of debate, and the idea that the right to a fair hearing belongs to everyone—with no exceptions. "Anyone can find fault" with Stalin, Dolutsky said. "And so I made the decision to be Stalin's defense attorney. . . ."

10
Paths
Not Taken

Wile *gulag* survivors have begun to tell their stories, and researchers have explored mass graves and labor camp ruins, historians and writers have been arguing about the causes. Why did those years after the Russian Revolution that were supposed to usher in the radiant future for all humankind turn, so quickly, into such a disaster?

For other mass murders, the explanations seem simpler. The white settlers of the nineteenth century who killed American Indians and natives of Africa wanted land. The Nazis who killed Jews wanted scapegoats—for the losses of World War I, the humiliation of the Versailles Treaty, and ruinous inflation and unemployment. But the Soviet Union's rapid devouring of 20 million or so of its own citizens remains more mysterious. The scale is so vast, the grievances so ill-defined, the accused villains so omnipresent, and the madness so extreme: "We must execute not only the guilty," said Nikolai Krylenko, Lenin's Commissar of Justice. "Execution of the innocent will impress the masses even more." Indeed it would. But where does this kind of thinking come from?

Historians in the United States and Europe, of course, have had decades to argue over such questions. But among Russians, to-day's discussion has a pent-up intensity to it, because for hundreds of years, scholars could not debate in full freedom. In Russia the historian's profession has never been an easy one. "God . . . merely decrees the future," Prince Kozlovsky said in 1839, "the Tsar can remake the past." Or, as one Soviet historian commented more recently, "The trouble is, you never know what's going to happen yesterday."

With the coming of *glasnost,* historians at last were able to speak openly about what did happen yesterday. One woman I met remembered the excitement of the first such talks, in 1987: "All Moscow was there in that small, miserable room. We fought our way through crowds, crawled under vehicles, to get to [historian Yu. S.] Borisov's lectures on Stalin. It was March or April, and it was still cold. Some journalists jumped in through the windows. People were outside, perched on the roofs of cars and on boxes, looking in through the windows."

Now that Russians can argue freely, they have staked out several major positions in the debate over the country's history. The crispest summary of them I heard came from Nikita Okhotin, head of historical research for Memorial. We were talking in a tiny, unheated office next to a room heaped with dusty piles of old newspapers, books, and documents, an apt symbol of a history that had not yet been sorted out—the makings of what will become Memorial's Moscow library.

"Right now," Okhotin said, "we have a set of myths about our past. The most widespread myth is the Solzhenitsyn myth. That is, Russian history was the history of an outstanding country, a great country, which suddenly, in 1917, experienced a disaster. Then there are variations of this: The Bolsheviks are to blame; the Jew-Masons are to blame.

"A second position is that things were always pretty bad. We never understood the idea of law. We didn't have a genuine intellectual culture, only a small island of people at the top. We always had authoritarian rule. The Revolution was not the destruction of everything; instead, it brought some trends in pre-revolutionary development to fruition.

"A final position is the *if* position. When Stalin took all power, everything changed for the worse. But everything would have been okay *if* Bukharin had been the man. Or everything would have been okay *if* Trotsky had been the man. The Revolution was necessary, inevitable, conceived correctly, but later on things went wrong."

Today, the first two positions in this debate sound through the country most loudly. For Russians, the tension between them carries an emotional significance far beyond its intellectual one. In the end, it is a debate about responsibility. Were the *gulag,* the Purge, the collectivization famine, and all the rest of it things that some alien, outside force—Marxism, Jews, Bolsheviks—did to us? Or was it something that we—Russians living out many centuries of Russian history—did to ourselves? It is a debate, finally, about whether Russians should see themselves only as victims, or as both victims and executioners.

The first position, which basically blames everything wrong on the Revolution (and, sometimes, the Jews), is the most popular. It is simple; it fits with the great popular revulsion against the years of Communist Party rule; it suggests convenient villains for the country's troubles. In general, however, I find more wisdom in the second school of thought, the one that stresses the continuities between Russia before and after the Revolution. One of the most articulate defenders of this point of view is Victor Doroshenko, a historian I talked to at the university in Novosibirsk:

"Stalinism did indeed change life in this country substantially, and some changes were horrible. But it was not an abrupt, dramatic change. People are now debating whether we should reintroduce private property. But there was none in the seventeenth century! Relations between people were *never* dependent on the market, but on one person obeying the other. This pyramid of power existed in the fifteenth century, and it exists in the twentieth. Peter the Great would gather people from different walks of life— the boyar, the merchants, the free peasants—and say to them, 'I need ships. So you chip in. One ship from five boyars, one ship from fifteen merchants, one ship from one hundred peasants.' So whose property did he commandeer? No one's property! All property belonged to the Tsar. What triumphed because of the Revolution? The idea of the property of the state, which has always existed in Russia over the centuries as the property of the Tsar.

"When our well-wishers abroad think that Russia is about to take a bright new road and begin living like a normal, civilized state, I can only say: Don't clap too hard. It's too early. If a society hasn't got the right way of thinking, what kind of rule of law can you establish? Haile Selassie asked René David, a French comparative law expert, to write a rule-of-law constitution for Ethiopia. René David wrote one. And what happened? Where is Ethiopia now? Where is Ethiopian law?"

The third way of looking at modern Russian history—the "*if*" position—is the most fanciful, but also the most intriguing. It allows us to think about turning points from which a different course might have resulted not in Utopia, to be sure, but in a very different Russia indeed.

Immediately after the Revolution, in the first national free election the country had ever known, the Bolsheviks won only a quarter of the vote. After the Constituent Assembly met for a single day, Bolshevik armed guards broke it up. What if they had not done so? Could some form of democracy, however flawed, have emerged? Three years later, rebel sailors at the island naval base of Kronstadt demanded freedom of speech and assembly and an end to special privileges for the Communist Party. What if Trotsky and Lenin had not sent fifty thousand troops across the winter ice of the Baltic to shoot them down? In the early 1930s, Soviet rulers combined peasants' plots into collective farms at gunpoint, triggering one of the century's greatest famines. What if they had heeded Trotsky's warning that this made as little sense as trying to combine a collection of small boats into an ocean liner? Each of these things would have been only one small step down an alternative path. But a Russia that took almost any such step would surely have come out less drenched in blood than it did. A whole range of "if's" shimmer, tantalizingly, on the receding historical horizon.

A sense of these possibilities animates Mikhail Shatrov's play,

Onwards . . . Onwards . . . Onwards! Shatrov himself is the nephew of a prominent Old Bolshevik ordered shot by Stalin; some of his work had to wait more than twenty years, until *glasnost,* to be staged. In this play, all the major characters of the early twentieth-century Communist movement step back and forth between scenes where they are planning the Revolution of 1917 and the next world, where they trade accusations about what went wrong. Rosa Luxemburg says that any revolution that suppresses freedom of speech is doomed to disaster. Kamenev tells Lenin of all that happened after his death. A Tsarist general says the real problem was the long delay in the nineteenth century in emancipating the serfs: "It came too late. If it had happened twenty years earlier, we would have been part of Europe."

Just as Christianity gave birth to both the Inquisition and the Quakers, Marxism, too, was a broad church. Politically, in Western Europe Marxism evolved into the Social Democratic parties of people like Willy Brandt and Olof Palme. Intellectually, shorn of its millenarian predictions, Western Marxism evolved into an analytic perspective whose sense of class is useful for understanding many things—including the enduring power of the old Party bureaucracy in Russia today. Was it inevitable that Marxism's Soviet branch would lead only to the *gulag?*

"It is often said," writes the novelist Victor Serge, "that 'the germ of all Stalinism was in Bolshevism at its beginning.' Well, I have no objection. Only, Bolshevism also contained . . . a mass of other germs—and those who lived through the . . . revolution ought not to forget it. To judge the living man [only] by the death germs which the autopsy reveals in a corpse . . . is this very sensible?"

In some of those who made the Russian Revolution—Stalin and the other ruthless sadists like Beria who rose with him—we can see the death germ, that fatal drive for absolute power. The finger on the map: I want a railroad—*there.* But in the character of other men and women who fought for the Revolution, we can glimpse a different set of motives, and an imagined Russia far different from Stalin's. It is especially poignant to follow the lives of these people in the 1920s, the point when it became clear their great dream of the transformation of human character was turning into a catastrophe. In the words of Serge, their most eloquent survivor:

> If we roused the peoples and made the continents quake,
> ... began to make everything anew with these dirty old stones,
> these tired hands, and the meager souls that were left us,
> it was not in order to haggle with you now,
> sad revolution, our mother, our child, our flesh,
> our decapitated dawn, our night with its stars askew ...

In the gloom of the 1920s, one of the most attractive figures in the shrinking circles of the anti-Stalinist opposition was Adolph Joffe. Joffe came from a wealthy Jewish family and was trained as a doctor, educated before the Revolution in Switzerland and Germany. He donated his entire inheritance to the revolutionary movement. Pictures from the time show him with a broad forehead and an air of thoughtfulness and dignity. "He ... reminded one of a wise physician ... who had been summoned to the bedside of a dying patient," wrote his friend Victor Serge. Before the Revolution, Joffe was a prolific writer, who worked with Trotsky in Vienna editing *Pravda,* which was then smuggled into Russia across the border or by leftist sailors on freighters crossing the Black Sea. The generation of revolutionaries Joffe came from was one whose passion for changing the world was matched by their free-ranging curiosity and by a nineteenth-century optimism that humanity stood on the verge of great breakthroughs in many fields. In Vienna, before World War I, he underwent psychoanalysis by Freud's disciple, Alfred Adler.

Joffe then returned to Russia to work underground, was arrested by the Tsar's police, and spent time in prison and in Siberian exile. After the Bolshevik takeover in 1917, he became one of the most prominent Soviet diplomats of his day, leading the first delegation to the peace talks with Germany at Brest Litovsk that ended Russia's participation in World War I. Soon after, as Soviet Ambassador to Berlin, he simultaneously represented his country to Kaiser Wilhelm II, and, as a revolutionary, flew the Red flag from his house and tried to encourage the revolts brewing among war-weary German workers.

Joffe became appalled at the direction Russia took in the 1920s. In place of the long-dreamed-of society free of class privileges was the tightening grip of the Party and the police; in place of a Utopia of free intellectual inquiry, forbidden books were withdrawn from libraries (Lenin's wife, Nadezhda Krupskaya, helped compose the first list); in place of debate was a press that increasingly deified Stalin. In 1927, at the age of forty-seven, Joffe returned to Moscow from his ambassadorial post in Japan, seriously ill. "In Moscow," writes the historian Isaac Deutscher, "doctors held out no hope for him and urged him to take a cure abroad. . . . Joffe asked to be allowed to leave at his own expense. Stalin then forbade him to publish memoirs, refused him an exit permit, deprived him of medical assistance, and harassed him with every kind of vexation."

Too ill to remain active in the opposition, unable to get medical help abroad, and wanting to express his horror at the disaster overtaking his country, Joffe killed himself. Victor Serge went immediately to Joffe's Moscow home when he heard the news:

> Joffe lay outstretched on a large table. . . . He slept, his hands placed together, his forehead bare, his graying beard neatly combed. His eyelids were tinged with blue, his lips dark. In the small black-edged hole in his temple, someone had stuffed a plug of cotton wool. Forty-seven years—prisons, the revolt of the fleet in 1905, Siberia, escapes, exile, Congresses, Brest Litovsk, the German Revolution, the Chinese Revolution, embassies, Tokyo, Vienna. . . . Nearby . . . a little room full of children's toys. . . .

On his bedside table, Joffe left a lengthy suicide note, in its way one of the more memorable documents of that sad twilight era. It begins by affirming the right to commit suicide. Then it expresses an all-encompassing humanism, which, though perhaps too tinged with technological optimism for our own time, affirms a larger faith.

... Thirty years ago, I adopted the philosophy that human life has no meaning except in so far as it exists in the service of something infinite—which for us is humanity. To work for any finite purpose—and everything else is finite—is meaningless. If one believes, as I do, in progress, one may assume that when the time comes for our planet to vanish, humanity will long before that have found the means to migrate and settle on other younger planets.... Thus anything accomplished in our time for humanity's benefit will in some way survive into future ages; and through this our existence acquires the only significance it can possess....

My death is a gesture of protest against those who have reduced the Party to such a condition that it is totally incapable of reacting against this disgrace [Stalin's expulsion of Trotsky and Zinoviev from the Central Committee]. Perhaps these two events, the great one and the little [Joffe's suicide], in occurring together, will reawaken the Party and halt it on the path that leads to Thermidor....

The authorities quickly declared Joffe's suicide note an illegal document; anyone found with a copy could be imprisoned. "The country at large," writes Serge, "did not hear Joffe's pistol-shot, and his last message remained secret."

Joffe's funeral on November 19, 1927, was the scene of an outpouring of support for the beleaguered anti-Stalinist opposition. A somber crowd of several thousand followed the coffin through the streets of Moscow to the Novodevichy monastery. Peter the Great had once imprisoned his own sister there; the monastery's beautiful, wooded cemetery is the final resting place of thousands of well-known Russians, from Gogol to Chekhov. (Remarkably, the mourners in the procession included Stalin's second wife, Nadezhda Alliluyeva, who was to commit suicide herself after an argument with her husband a few years later.) Secret policemen roughed up the crowd and tried to disperse it, but there were too many people. Trotsky spoke at the graveside, the last public speech he was allowed to give in Russia. A few weeks later, he was sent off to exile in Kazakhstan. Over the next few years anyone who dared criticize the regime openly was immediately jailed or shot. Joffe's funeral was the requiem for an entire era—the period when there still seemed a slight, diminishing hope of an alternative to dictator-ship.

The crowd at Joffe's graveside did not know it, but theirs was to be the last mass demonstration in Moscow in opposition to the government for some sixty years.

A gray-haired woman in a white sweater sits in an armchair in her Moscow apartment, the spring sun streaming through the windows. At eighty-five, her eyes are vibrant with humor. She vividly remembers the funeral procession that day, for she is Nadezhda Joffe, Adolph Joffe's daughter.

She remembers another day, too, ten years earlier: November 7, 1917. On that momentous evening, and throughout the night, Adolph Joffe and his wife were at the Smolny Institute in Petrograd—Party headquarters and the nerve center of the coup that installed the Soviets in power. It was from the Smolny that orders went out for armed Party militants to seize the Winter Palace, the telephone exchange, the State Bank, the railway stations, and other key points around the capital. And it was from the Smolny that the new revolutionary regime proclaimed its existence to the world.

On that night that changed the course of twentieth-century history, the young Nadezhda Joffe was also at the Smolny Institute—her parents couldn't find a baby-sitter.

She remembers her father warmly. "When I was twelve, we were in the embassy in Berlin, when Father was ambassador. There I began to read the press, both Russian and German. I got interested and asked my father the most stupid questions imaginable. But he never said things like, 'You're still a kid, you can't understand that.' He'd always answer all my questions."

Two years after her father's death, pregnant with her first child, Nadezhda Joffe was arrested for the first time. She was to spend more than twenty of the next thirty years in prison, labor camp, or exile. But she was lucky, she says, laughing. "When I was jailed the first time [1929], they weren't beating people yet."

Returning to Moscow in the thirties after her first Siberian exile, she was appalled at the wave of "capitulations" by prominent Old Bolsheviks, who now were falling over themselves to confess their "errors" and beg for readmission to the Party. Times had changed since her father fought against Stalin a decade earlier. "People were leaving the opposition. They were making those statements. Famous people. People like me, young people, who were not so important, they simply wrote: 'Please add my signature to such-and-such a statement.'" Without a "capitulation" statement, the authorities wouldn't give her a residence permit. "We had two kids. Everybody was after me. 'You should think about the children! You should think about their lives,' urged my friends, who'd once been in the opposition, but by that time had written their statements. I didn't."

In 1936, Joffe and her husband were again arrested; this time they were sent to the furthest and coldest corner of the *gulag,* the territory whose name became a synonym for the ultimate in Stalinist horror, Kolyma. In a labor camp hospital there, Joffe gave birth to another child, who was taken away to an orphanage for children of "enemies of the people." Her husband was shot.

Joffe was sent to a remote gold mine. In Kolyma, where malnourished prisoners had to face the ferocious Arctic cold without adequate clothing, time was often the equivalent of a death sentence. She survived because she had a job in the kitchen—and because the leader of a criminal gang decided to protect her. "His name was Sasha Orlov, but his nickname was the Eagle. He was older than me, but he always called me Nadezhda Adolphovna. He was very respectful. He would talk and talk, and then would say: 'You know, I'm surprised at myself. I've had more girlfriends than hairs on my head! I normally wouldn't talk to a broad as if she were human. But I'm talking with you!'" One day one of the other common criminals broke into the barracks room where Joffe lived and stole her jacket and a bottle of cologne. "Sasha came and asked, 'Why are you so sad?' I explained. He said, 'Wait a few minutes. I'll be back.' Half an hour later he was back, and brought the jacket. He said, 'Sorry, they've drunk the cologne. But from now on your things will be safe. These are the orders of the Eagle.' "

Joffe was released to "the mainland" from Kolyma in 1946. In 1949, when another wave of arrests began, she was imprisoned again for almost a year, then sent into exile in a small town near Krasnoyarsk—the place of her first Siberian exile some twenty years earlier. Her children were again scattered to relatives and orphanages. She remained in exile until she was rehabilitated in 1956. Today she is preparing to emigrate to New York, to spend her remaining years with a daughter there whom she has visited once or twice.

Has she changed her view of the world over the years?

"You would be extremely stupid not to. But it was very difficult for me. My youth was a wonderful time, a time of absolute commitment. I didn't need anything for myself, we needed only world revolution. I have an American friend living in New York, N——. Do you know him? I stayed at his place in Manhattan. What can I tell you?—you know that life better than I. He has a nine-room apartment: three bedrooms, two bathrooms, everything. But he doesn't want it. He wants world revolution! But I love him dearly. He's such a sweet person. I, too, was brought up with all those beliefs."

Joffe describes several similar encounters with Western leftists—one with an American Trotskyist in Moscow, one with the actress Vanessa Redgrave. She tells each story not with the contempt most Russians have for Western radicals, but with a wry compassion, as if she were recounting friendly arguments with some earlier version of herself. She has not given up on Utopia; she still empathizes with these leftists' passion for social justice. But what bemuses her so much is to find it among people living in the United States and England, countries that look, from Russia, like paradise.

I ask Joffe what she believes in now, and her answer recalls her father's last letter, searching for the infinite in a finite world. "I don't believe in God. But people I really respect have turned to religion. I'm such a rational person: I took the Gospels and I read all four. Luke, John, Matthew . . . And I liked all these fine beliefs: you should not kill, should not steal, should respect your mother and father. I try to stick to those beliefs. But I don't believe in God, either a Russian Orthodox one or a Jewish one. None."

At the end of our meeting, she tells me a joke; like a haiku verse, it contains a hint of what went wrong in the Russian Revolution, and of what might have been, a path not taken.

"A descendant of the Decembrists [reformer aristocrats who rebelled against the Tsar in the 1820s], a woman, sees a crowd demonstrating in the street and she sends her daughter outside: 'Masha! Go and see what's going on.'

"Masha returns and says, 'Lots of people are out on the street.'

" 'What do they want?'

" 'They're demanding that no one should be rich.'

" 'That's strange,' says the woman. 'My grandfather went out onto the street and demanded that no one should be poor. . . .' "

11

"The
Interrogation Has
Been Interrupted"

Despite all the changes brought by *glasnost,* at the time I was in Moscow, one significant part of the national memory remained almost totally hidden: the records of the executioners. Survivors have told us about their experience of interrogation, torture, and prison, but what would the notes of the jailors themselves say? It was nearly impossible to find out.

The major reason is that though much has changed in Russia, the secret police are still there, working in the same rooms and buildings where people were tortured and shot some forty years ago. The bulk of the *gulag* was not shut down until 1956, three years after Stalin's death, and some older officers still now in uniform began their careers earlier.

Like many a product trying to improve its image, the Soviet secret police has changed its name at various times over the years; parts of it have been spun off to other ministries and then taken back again. If you draw a chart of all these changes, it begins to look like a genealogical diagram of a large, heavily intermarried, divorce-ridden, extended family. Immediately after the Revolution, the secret police was known as the Cheka, an acronym from several words in its full title: the All-Russian Extraordinary Commission for the Struggle Against Counterrevolution, Speculation and Sabotage. It changed its name twice in the 1920s, and then in 1934 became the NKVD—the People's Commissariat of Internal Affairs. After several more divisions and name changes, the core of the force became the KGB, or Committee for State Security, in 1954. But officers still proudly call themselves *Chekisti,* just as they did seventy-five years ago.

More recently, as the Soviet Union dissolved, there has been yet another layer of branches and name changes in each republic. In Russia itself, the KGB has already changed its name twice just since the coup attempt of August 1991, though everyone still calls it the KGB.

The great waves of mass arrests and executions ended after Stalin's death, yet the force remained immensely powerful. The KGB persecuted and exiled people like Solzhenitsyn and Sakharov, and sent thousands of dissidents to jails, labor camps, and psychiatric hospitals. Visiting the Soviet Union during the 1960s, 1970s, and early 1980s, I often found that whenever conversation became too political, people would point meaningfully at an apartment's ceiling, and, instead of talking, would start scribbling notes to each other, which they later burned in ashtrays or tore up and flushed down the toilet.

By the beginning of the 1990s all this had changed dramatically. But without listening in on hidden microphones and the like, what's a secret policeman supposed to do all day? The KGB was in transition—but to exactly what, no one knew. The government several times announced that the agency would play a key role in the country's battle against rapidly growing organized crime. But, just as J. Edgar Hoover's FBI long found snooping on America's minuscule number of Communists much simpler than pursuing the Mafia, so the KGB found harassing the pre-*glasnost* tiny community of dissidents far easier than pursuing real criminals. By the time I got to Moscow, rival gangs from the Caucasus extorted merchants and restaurants by day, and by night took to the city's parks with knives and guns to settle scores with each other.

As reforms went forward, KGB officials worried that the agency might be shrunken or dismembered. For now that censorship had dissolved, adding to the KGB's woes was a bad press. To fight back, the agency started publishing a magazine, set up a small museum in its headquarters at the Lubyanka building near Red Square, announced production of a joint Soviet-Italian thirteen-episode TV series called *The KGB Tells All,* and held a "Miss KGB" contest, in which the contestants wore bulletproof vests. The winner, Ekaterina Maiorova, declared her favorite hobby knitting, her favorite drink orange juice (a nod to Gorbachev's ill-fated temperance program), and her favorite film *Gone With the Wind.* Her favorite man was James Bond.

To get a taste of the KGB's new look, I went with some friends one day to an unusual forum in a public meeting hall, advertised as "The KGB Without Mysteries." This was billed as the first meeting at which KGB officials would answer questions from the general public.

The public, however, did not look too general. Most seats in the packed auditorium were occupied by stolid-looking men in their thirties and forties who seemed to have arrived early. Other chunky young men were stationed at the doors, their eyes roaming the crowd. A more ordinary-looking variety of Muscovites of both sexes filled the aisles and stood around the back and sides of the room.

Miss KGB, alas, was nowhere in sight. Instead, the dozen or so people on stage were all older men, mostly on the heavy side and looking ill at ease. Around the walls of the auditorium was a photo exhibit on the Border Guards, a unit of the KGB that defends "the gates of the motherland" with helicopters, ships, guard dogs, and snowmobiles.

Whatever their faults, KGB officers don't speak for more than their allotted time. Each man took his ten minutes and then promptly turned over the microphone to the next. After each speaker, the men in the auditorium's center seats applauded dutifully. The crowd standing in back or sitting in the aisles was mostly silent.

Most speakers echoed a familiar theme: the source of all trouble in Russia is foreigners. Colonel Anatoly Grinenko of counterintelligence, for example, charged that Americans and Europeans had forged connections with organized crime. "We can expect Western intelligence agencies to try to exploit these links for their own purposes."

Another speaker was apparently an in-house novelist. Ivan Shevtsov was introduced as a former Border Guard now writing on "the Border Guard theme." He said: "Even in the first hours of the war, we believed in our victory. We believed in the values that are now trampled and spat upon!" In the question period, one member of the audience accused Shevtsov of being anti-Semitic. The novelist indignantly defended himself by naming various of his villains who were *not* Jewish.

All in all, the KGB's new enlightened look was not much in evidence. But I was less interested in the KGB today than in its vast archives, and the secrets they might hold about how the Russian Revolution had turned on itself. Because the secret police archives were still essentially closed, even people who were rehabilitated seldom saw their own files. Usually, they were only sent a certificate saying that they were not guilty after all. Although this was later to change, at the time of my stay in the city in the first half of 1991, none of the Memorial researchers I met had ever set foot in the KGB archives in Moscow. And if the KGB was not eager to share its materials with Russian researchers, I thought, it surely was not going to show them to a foreigner.

I was wrong.

One May afternoon, as I was making preparations for my long-planned trip to Kolyma, the telephone rang. The caller was another American I had gotten to know in Moscow, Kathleen Smith. She was writing a Ph.D. dissertation about Memorial, and we had occasionally interviewed people together. Smith had just met a woman from the Moscow prosecutor's office, she said, who had a stunning piece of news: "The KGB has some files about Americans in the *gulag*. They're looking for an American journalist to give them to. And they'll give us a tour of their archives."

On a sunny morning a few days later, I met Smith on the

square outside the Bolshoi Theater. It was here that her contact in the prosecutor's office had arranged to meet us and take us to the KGB. The last time I'd gotten off the subway at this stop, the station had been called Sverdlov Square, after an early Bolshevik leader. Now, I noticed, as part of the countrywide epidemic of name changes, it had just been renamed Teatralnaya, after the theater.

The square was an ironic spot for us to meet, for it is the site of an annual summer reunion of old *Kolymyaki*—survivors of the Kolyma labor camps who live in Moscow. But Arctic Kolyma felt very far away on this warm spring day when the trees were all in leaf. After waiting for a few minutes on a bench outside the theater, Smith and I were joined by Olga Matlash, a pleasant, forthright woman in her mid-thirties. She was the supervisor of rehabilitations in the Moscow office of the state prosecutor. It was this work that had led her to old secret police files on people with U.S. connections.

The Soviets began slowly rehabilitating Stalin's victims soon after his death in 1953. But the roll of names of those jailed or shot was some 20 million long, and rehabilitation officials got only a small part of the way through it. After he took power in 1964, the conservative Brezhnev quickly brought rehabilitations to a stop—and actually *un*rehabilitated some people whose names had just been cleared. More than twenty years later, Gorbachev started rehabilitations again in earnest, although by this time most victims of Stalinism were dead. The Russian government today claims to have more than two thousand employees working on rehabilitations.

Rehabilitation seems odd to us because, as a secular society, we put little stock in posthumous status. But Soviet communism from the beginning was, psychologically, a religious culture. Just as the Great Purge was Inquisitorial in its fervor, so rehabilitation, and posthumous restoration of Party membership, have offered a kind of sainthood for martyrs. (The Inquisition also operated posthumously: the dead could still be convicted of heresy, and their bones burned.) For Soviet prison camp survivors still alive, rehabilitation is also the nearest thing to an official apology. Many times in my six months in Russia, people showed me rehabilitation certificates for themselves, or for a mother or father who had been shot. There is a practical side to all this, too, increasingly urgent in a collapsing economy where the elderly suffer most: rehabilitated people still living get compensation payments (the sum is based on how long you were in prison), and sometimes, like World War II veterans, extra food and pension benefits.

After a short walk, Olga Matlash, Kathleen Smith, and I arrived at attractive old building of blue stucco with gingerbread white trim. It had been a pre-revolutionary mansion; from the balcony, Matlash told us, Napoleon had watched Moscow burn.

A small plaque with gold lettering on black plastic, the same style as for all Soviet government agencies, marked the building as the *priom,* or reception room, of the Moscow region of the KGB. The prosecutor's office has "oversight" over the KGB, Matlash explained. But she was so nervous as we arrived at the KGB *priom* and then waited on the sidewalk outside until exactly the time of our appointment that the "oversight" seemed the other way around. Promptly at 10:00 A.M., a thin, impassive man in a pin-striped suit and blue tie came out onto the sidewalk, introduced himself as Colonel Nikolai Grashoven, and conducted us inside. He took us past a raised counter manned by a uniformed guard, then through several doors fitted with remote-controlled, electrically operated locks, with black wires running to them along the walls.

We found ourselves in a bare inner office with a wooden desk and adjoining table and chairs. On the desk were three yellow telephones. These rang periodically during the next hour or two, often with wrong number calls; even the KGB, it seemed, was not immune from the ancient switching equipment that sends so many Moscow phone calls astray. A white curtain covered a window. The walls were paneled in mottled birch; on one was a photograph of Felix Dzerzhinsky, the founder of the secret police. In mustache and goatee, Dzerzhinsky looked shrewd and almost rakish, gazing down at us out of the corner of his eye as we took our seats around the table.

Colonel Grashoven was forty-nine years old, with pale, washed-out blue eyes. He was the chief, he said, of a group of thirty KGB officers in the Moscow area who work on rehabilitations, which means that he is constantly dealing with Stalin-era archives. In the three hours we spent in this room, on this day and the next, Colonel Grashoven did most of the talking. His face was angular and forbidding; guilty or innocent, I would not want to be facing him on the other side of an interrogation table. However, in contrast to many other Soviet officials I've talked to over the years, he had a no-nonsense military directness. He made clear the reason we had been invited here—public relations.

"We're interested in showing you [Americans], through your media, that we're not withholding information from you about what we're doing," the colonel said briskly but without enthusiasm. Perhaps he was overcoming some repugnance at the idea of showing a lot of old dirty linen to foreigners, but orders were orders. As part of the KGB's desire to show how open things are now, he continued, the agency was releasing, to one correspondent from each major Western country, information about people from that country who disappeared in the Soviet Union during the Stalin years.

Americans and Europeans inquiring about vanished relatives are now encouraged to write for information. The colonel estimated that the KGB has files on "several thousand" Stalin-era victims who either were born in the United States, lived in the United States for a time, or have relatives there today. As proof of the agency's new openness, he went on, they would give us some sample files.

After a few minutes' talk, Colonel Grashoven's deputy entered the room, carrying a bundle of file folders under his arm. Major Mikhail Kirillin was thirty-eight, spoke with a Byelorussian accent, and had a thick shock of black hair and a mobile face that went quickly to a smile or frown. He had recently helped find information for French and Italian groups of relatives of Stalin's victims. Because of this, he told us with great animation, he had been invited to Paris the next month. It was the first time, he joked, that any KGB officer had gone there "by invitation." Several times dur-

From left: Major Mikhail Kirillin, prosecutor Olga Matlash, and Colonel Nikolai Grashoven, in the KGB archives.

ing the next hour, the subject of Major Kirillin's scheduled trip to Paris came up, and each time his face uncontrollably broke into a delighted grin. I couldn't help liking him.

Before getting down to business, the two officers seemed eager to talk about the KGB itself and especially about its long-time chief, Yuri Andropov, who went on to head the Soviet Union briefly before his death in 1984. "In a way," Colonel Grashoven began, "the KGB triggered *perestroika,* telling the country's leaders the truth about what was going on—in the economy, in transportation, in agriculture."

"I'd like to tell you something about *perestroika,*" added Major Kirillin earnestly. "In the West, they count its days from 1985 [when Gorbachev took over]. But we *Chekisti* count it from 1982, from when Andropov came to power. When the government was made up of seventy-five-year-olds, they didn't realize what the real situation in the country was. The apparatus that surrounded them met their needs, and they played their game. But Yuri Vladimirovich [Andropov], having spent fifteen years in this organization, knew the situation in the country from A to Z." Echoing the assessment made by many others, he added, "Without Andropov there would have been no Gorbachev."

Although the two officers acknowledged that the KGB had unjustly harassed dissidents in the pre-Gorbachev years, they claimed that even then 98 or 99 percent of its manpower was at work at legitimate tasks like catching common criminals. (Who, then, was keeping track of "the situation in the country from A to Z"? I didn't press the point.) But they did not try to downplay the secret police role in Stalin's terror. At one point, referring to the KGB's national headquarters just down the block, Colonel Grashoven even made the familiar joke: "What's the highest building in Moscow? It's the Lubyanka. Because from the top floor, you can see all the way to Kolyma...."

Listening to the two men talk, looking like any other bureaucrats in their suits and ties, I wondered whether the rehabilitations section of the KGB was a career-path backwater—a distinct feeling I had had once when I happened to meet these officers' closest American counterpart: a dispirited CIA man who dealt with people looking for old files under the Freedom of Information Act. Surely, I hinted, rehabilitating the long-dead must feel less glamorous than catching Soviet crooks or American spies. Major Kirillin confessed that he had been dismayed when first assigned to this work. "I had been in on the beginning of the Uzbekistan case. I grabbed the hand giving the first bribes. Suddenly I was sent to the archives!" (The Uzbekistan case was a huge embezzlement scandal connected with cotton production. The hand Major Kirillin had grabbed, it seemed, may have belonged to someone with friends in high places.)

But now, both men maintained, they were glad to be here, and were flooded with applications for transfers from people in other KGB departments. The reason they gave sounded plausible: it's the only job in the KGB where people are grateful to you. Major Kirillin described one old woman who had been trying for years to find out what had happened to her father. They found his file. She wrote back, asking "that somebody come to her, because before her death she wants to look into the eyes of the person who had found the fate of her father and give him a big, human, peasant thank you."

"To be objective," Colonel Grashoven added, "it should be said that ninety-nine out of one hundred are letters of thanks, but there's always one: 'Get lost! For God's sake, we've forgotten everything and you're reminding us!' Or, 'I spent twenty years in the camps . . . get me an apartment! Give me a washing machine!' People are sure that the KGB can do anything."

Sometimes, said the major, "we invite someone who'd been repressed to come in and pick up his rehabilitation certificate, and he says, 'Anywhere else, just not in your building . . .'"

Unfailingly polite, the colonel and the major were among the few people I met in Russia who didn't ask, "And why are *you* so interested in all this?" Instead, eager to show their openness, they seemed ready to talk about whatever we wanted. Kathleen Smith and I brought up a question historians have long argued about: Just how many deaths was Stalin responsible for? The two officers referred to some of the different positions in this debate, which they seemed very familiar with, but they did not take one themselves. Here in an adjoining building, they said, they have records of 120,000 people shot or sent to the *gulag,* but these are only half the existing records for the Moscow area alone. Many more files have been lost or destroyed.

During this conversation, it was often hard to remember that I was talking to two secret police officers. Except for their defense of the present-day KGB, almost everything Colonel Grashoven and Major Kirillin had to say could have come from one of the Memorial activists I had interviewed over the last few months. The major knew Shalamov's prison camp short stories; the colonel quoted the poet Nekrasov about Russian passivity and Alexander Herzen on the nature of guilt. They used words like "totalitarianism" and "slave labor." Colonel Grashoven ruefully blamed "the Iron Curtain—nobody in, nobody out" for the fact that the country was so backward that his rehabilitations department didn't even have a computer. Echoing many Russians I had talked to, Major Kirillin said it would take two generations for the bureaucracy to really outgrow its old authoritarian habits. He talked about the reason so many of the best intellectuals left the Soviet Union in the 1920s: "Why? Because any reasonable person, reading Article 58 [of the criminal code, which had vague and sweeping prohibitions against "counterrevolutionary" activity] would understand everything. A person could be arrested for *anything.*"

Was all this merely a performance for my benefit? Had the two officers been carefully trained to say what a Western writer would want to hear? Possibly, but I doubted it. They lacked the practiced veneer of many Soviet officials who spend a lot of time talking to foreigners, and they did not speak English.

The other possibility was that they spoke what they believed. Like other parts of the Soviet power structure, the KGB long had in its ranks both hard-liners and those who knew that some kind of drastic change was necessary. Given what the Soviet secret police has represented over the years, it would be tempting to lean toward the more cynical explanation of the two officers' opinions. But the more I think back over this conversation, and listen again to my tape recording of it, the more I hear in their voices the ring of a certain sincerity.

After we had been talking an hour or so, we began looking through the stack of papers Major Kirillin had brought into the room with him. They were in file folders of faded purple or tan cardboard, each one labeled "Case Number—" in large black type. These were five cases, found by Major Kirillin in the archives, of *gulag* victims with American ties. Each folder was thick with yellowing, dusty paper: transcripts of interrogations, forms listing confiscated belongings, indictments, sentences, and death certificates, all typed on the manual typewriters of the day or filled out by hand. None of the prisoners whose fates were recorded in all this paperwork was at all prominent, but the files on them were large, the longest, fifty-four pages.

Stalin's men were not alone in keeping meticulous records. The military rulers of Brazil did so also, which later embarrassed them when it provided evidence of widespread, systematic torture. The Nazis kept enormous amounts of paperwork as well. "Procedure meant a great deal to our rulers," writes Nadezhda Mandelstam, "and the whole farrago of nonsense was always meticulously committed to paper. Did they really think that posterity, going through these records, would believe them just as blindly as their crazed contemporaries?" Now, in this room, for these particular records, we were the first representatives of "posterity."

In these files opened on the table before me, an occasional page seemed to be missing. But in general, the cases were amazingly complete. I had expected that, as in CIA and FBI files released under our Freedom of Information Act, names of police agents and informants would be whited out. Not so. Everything was there: the names of denouncers, of interrogators, of their superiors, of secret police units, and of the victims' anguished relatives, begging for information. In each file of dusty, brittle pieces of paper lay a mosaic of an entire life, and sometimes of a whole family's life as well.

Two cases stand out with particular vividness. One of these young men was born in New York, the other in Massachusetts. In Moscow, they moved in different circles and probably did not know each other. But by coincidence or plan, both were shot on the same day in 1938. And possibly by the same executioner: in each file, the brief certificate stating that the death sentence has been "implemented" is signed by the same scrawling hand.

In the first file, an NKVD biographical information sheet lays out the facts. They are filled in by hand, in blank spaces like those of an application form:

> *Name:* Arthur Talent
> *Place of birth:* Boston, USA
> *Year of birth:* 1916
> *Health:* good
> *Occupation:* artist

Army Service: none
Party Membership: none

Arthur Talent had been brought by his mother from the United States to Russia when he was six years old. He lived in Moscow for the rest of his short life. At the time of his arrest on January 29, 1938, he was twenty-one.

It was not a good moment to be arrested. That very day, a new chief, Leonid Zakovsky, took over the Moscow branch of the NKVD. According to Robert Conquest, Zakovsky immediately called a meeting of the city's top *Chekisti,* railed at them furiously for being too slow in rounding up "enemies of the people," and ordered a daily quota of two hundred arrests. This pace surpassed even Torquemada, who during the Inquisition managed a mere thirty-five prosecutions of Spanish heretics per day. Zakovsky actually had an eye for such historical comparisons: he reportedly boasted that with his techniques he could have made Karl Marx confess to being an agent of Bismarck.

Arthur Talent was held at Moscow's Taganskaya Prison. From survivors' accounts, we know that the cells there were dirty and dark, with paint peeling from the walls and floors of broken asphalt. Senior prison officials lived in an adjoining building; the commandant's flat had a piano.

One thing immediately sticks out in Arthur Talent's file: the officer who kept the record of his first interrogation, a Lieutenant Salov, was not very literate. Many commas and periods were missing from his transcript. In the twenties and thirties, the lower ranks of the secret police often came from the working class or the peasantry. Two interrogators grilled Arthur; some of the rage that they seem to have felt toward their prisoner may have been sheer class envy. Arthur Talent had been born abroad and apparently spoke at least three languages, English, Russian, and Latvian (his parents had immigrated to the United States from Latvia before he was born).

The warrant for Arthur Talent's arrest says only that he was sought for "involvement with espionage activity." His interrogators' questions reflect that ancient Russian suspicion of anything foreign. The first interrogation, two days after his arrest, begins:

QUESTION: When did you work in the Novomoskovskaya Hotel?

ANSWER: I worked as an elevator operator in the Novomoskovska-ya Hotel for five months in 1932.

QUESTION: What foreigners were your close friends when you worked in the Novomoskovskaya Hotel?

ANSWER: I met with a foreigner, a Negro, Henry Scott, who per-formed in the hotel's restaurant as a dancer.

Page after page, the interrogators ask Arthur about his contacts with foreigners. For a time, Arthur says, he shared an apartment with Henry Scott, who performed with a jazz band. Arthur met many of Scott's friends: a jazz musician, an American boxer, "Olsen, a correspondent for a foreign newspaper," an anonymous German, "a certain John Goode ... a foreigner and a Negro," and John Goode's sister, who was married to Paul Robeson, the singer.

After getting all these names from Arthur, the NKVD interro-gators ask him,

QUESTION: Judging from the above, what purpose do you think your apartment served?

ANSWER: I have to admit that my apartment was a clandestine ad-dress of a foreign state, and I was its keeper.

QUESTION: You have not told everything to the inquest. We de-mand you give true testimony.

Later the interrogator asks:

QUESTION: What state used your apartment as a clandestine ad-dress?

ANSWER: I am not able to give an exact answer to that question, since John Goode is an American citizen. Henry Scott also. While the person who sold me the boots is a German citizen. . . .

Soon after this, the transcript ends with the words: "The interrogation has been interrupted."

What went on, unrecorded, between Arthur Talent and whoever "interrupted" this interrogation? It is not hard to guess. By 1938, torture of Soviet prisoners had long been routine. Even the Purge's elaborate facade of legality didn't try to conceal this: NKVD regulations explicitly allowed "physical pressure" in questioning prisoners. Thomas Sgovio, a young American who had arrived in Moscow from Buffalo, New York, some three years earlier, was an inmate in Taganskaya Prison at exactly the same time as Arthur Talent. In the prison's communal baths, he writes, "We stood in packed lines of bony humans, more than half of them covered with blue, blood-crusted welts from beatings received during interrogations."

Arthur Talent probably was desperate to admit to something to save himself from these beatings. Kafka, in *The Trial*, had mysteriously anticipated all of this more than twenty years before. Another character tells the accused, Joseph K.: "You can't fight against this Court, you must confess to guilt. Make your confession at the first chance you get. . . ." But, like Kafka's hero, Arthur sounds bewildered. What was he supposed to confess to? Being a spy for the United States? For Germany?

The next interrogation picks up where the previous session left off. What the dry words of the dialogue do not show is the atmosphere in which it probably went on: interrogations were routinely done at night, with bright lights shining into the prisoner's eyes. Arthur, apparently more frightened than ever, now makes up something specific to confess to:

QUESTION: The inquest demands that you stop disavowing and tell how you were recruited for espionage activities.

ANSWER: I plead guilty that I have been concealing from the inquest that my apartment was a secret address of a foreign intelligence service. I learned this from John Goode, who was living in my apartment. . . .

QUESTION: When did John Goode tell you that your apartment was a clandestine address for foreign intelligence, and what state did it belong to?

ANSWER: It was at the end of 1936. John Goode . . . told me that my apartment was going to be a clandestine address for British intelligence and asked me not to tell anybody about it, and promised me a payment for being silent.

In naming John Goode as a British spy, Arthur must have hoped he would satisfy his interrogators' relentless hunger for names of "conspirators." Perhaps for a moment he did: the file shows that because of his testimony, the police issued an arrest warrant for Goode. But Arthur knew he was not putting his friend in any danger. According to books on the period, Goode had moved back to the United States the previous year. All Soviet prisoners faced the dilemma of what to do when asked to name names. One memoir comments: "The best way out was to name people who were dead or had left the Soviet Union for ever."

After this last exchange between Arthur and his questioners, an ominous gap appears in the file—five and a half weeks for which there are no interrogation transcripts. The next recorded interrogation, before a new officer, is dated March 8, 1938, thirty-eight days after that last session. We have no way of knowing what happened to Arthur during those thirty-eight days, but it probably was not pleasant. His signature, large and confident when he had signed the formal acknowledgment of his arrest, now becomes small and shaky, the letters cramped together, as he signs the transcripts of this new round of interrogations.

By contrast, the handwriting of Arthur's new interrogator is firm and bold, and it is strewn with underlinings and exclamation marks in the parts recording Arthur's confessions. It is as if we can look over the interrogator's shoulder and see him glorying in extracting these statements from the terrified prisoner in his charge.

These underlinings and exclamation marks also speak of the intense pressure on interrogators. "Those who could obtain [a confession] were to be considered successful operatives," writes Robert Conquest, "and a poor NKVD operative had a short life expectancy." Conquest quotes a manual for secret police officers of the thirties: "... Failure to confirm the evidence ... is indicative of poor work by the interrogators." Almost never, however, was there any evidence—but that just made confessions even more important.

The new interrogator seems to be under orders to get Arthur to confess to spying not for Britain, but for Latvia. In the handwritten record, the interrogator has vigorously underlined Arthur's answer to his first question:

QUESTION: You are arrested and accused of espionage activities in the USSR on behalf of a foreign state. Do you plead guilty?

ANSWER: *Yes! I plead guilty to spying for Latvia. After a thirty-eight-day disavowal I have decided to tell the inquest the whole truth.*

After now confessing to being a Latvian spy, Arthur is interrogated a second time the very same day by two more officers. One was a major, who seems to have stepped in to gain credit for this much-wanted confession. An NKVD major then was equivalent in rank to a Red Army brigadier general. Why should such a senior official bother with a twenty-one-year-old artist and former elevator operator when there were so many higher-ranking Great Purge victims to interrogate? We can only speculate. Finding Latvian spies might have been a good path to promotion in 1938; some months earlier, Stalin had ordered a sweeping purge of the Communist Party of Latvia. The NKVD dutifully responded by uncovering various Latvian conspiracies in the Red Army and elsewhere. At that time Latvia was an independent country, which the Soviet Union was preparing to absorb, via the Hitler-Stalin Pact of the following year. It may be ripples from these events that shape the major's interrogation of Arthur:

QUESTION: Who recruited you into this organization and when?

ANSWER: At the beginning of 1937, I was recruited into the Latvian counterrevolutionary espionage organization by Bantsan, brother of the director of the Latvian Theater.

QUESTION: What do you know about the goals of the counterrevolutionary Latvian nationalist espionage organization?

ANSWER: When Bantsan recruited me, he told me that the organization wanted to unite all Latvians to overthrow Soviet power and create a mighty Latvia on the territory of the Soviet Union.

Arthur Talent goes on to name several actors in the Latvian Theater in Moscow. Now we see a hint about why he had evidently held out for five and a half weeks before naming these people, who were apparently his friends. Unlike John Goode, safely in America, the Bantsan brothers and the Latvian actors were in Moscow. Thanks to the NKVD's careful cross-reference filing system, carbon copies of their documents are in Arthur Talent's file. Immediately after Arthur named them as Latvian spies, these papers show, both Bantsan brothers and one of the actors were arrested and shot.

Stalin's paranoia about conspiracies against him was never limited by probability or geography. Creating "a mighty Latvia on the territory of the Soviet Union" was about as feasible as creating a mighty Costa Rica on the territory of the United States—with a key spy ring strategically placed in a Costa Rican theater in New York.

Logic, however, was not the point. Stalin always imagined his enemies not as lone rebels but as members of plots and groups. If one person was arrested, he or she had to have co-conspirators. An interrogator who did not come up with such names might be accused of insufficient vigilance and shot. "It was the duty of a patriot," explains Nadezhda Mandelstam, "to fulfill his quota. Names given to an interrogator could either be used immediately or kept in reserve against a shortage. Names of 'accomplices' were the interrogator's stock in trade, and in case of need, he would dig into his reserves. . . . Even when the terror was at its height, some kind of nominal excuse was always found for a person's arrest: a denunciation, information from police spies, or best of all, the fact that your name was mentioned during the 'investigation' of someone else."

After supplying the necessary names, Arthur Talent was indicted. Just to be safe (those who drew up indictments could also be accused of insufficient vigilance), the indictment charges him with ties to both Latvian and British intelligence. As part of the evidence against him, it includes the accusation that he "led a wide-ranging life, often visited restaurants, and had connections with foreigners."

All the documents in the file are numbered, in order by date. The next piece of paper is small and crumpled. So many executions happened at this time that they were recorded on a special printed form, about the size of a traffic ticket. Only the prisoner's name and the dates are handwritten, filling in the blanks:

The Resolution of the NKVD of the USSR of May 23, 1938, on the shooting of Arthur Karlovich Talent was implemented on June 7, 1938.

A space for the time of day is left empty. The signature is illegible. Arthur Talent was shot two months after his twenty-second birthday.

For nearly two decades after Arthur Talent's execution, no more documents appear in his file. Then, in 1957, Arthur's maternal aunt, Minna Gabalina, appeals for information about her nephew's fate. From the whole extended family, it seems, she alone survived to tell its story.

Minna Gabalina describes her nephew as "a modest young fellow." She has more to say about her sister, Arthur's mother, Elena, who was arrested around the same time. The letter makes clear the real reason for Arthur's arrest: his parents were both Old Bolsheviks, Party members from before the Revolution. In her appeal, Minna Gabalina summarizes the family's history. It began in Latvia, which before World War I was part of the old Russian Empire. "In 1910," she writes about Arthur Talent's father, ". . . Party member Karl Veniaminovich Talent was arrested by the Tsarist police and sent to permanent exile in Yenisei province [in Siberia]. In 1911 he escaped. . . . He and his wife, E. M. Gabalina-Talent, an underground worker, escaped abroad, first to the city of Bordeaux in France and from there quickly to Boston in the USA." Her sister's code name in the Party underground was "Pigeon." Minna Gabalina herself helped smuggle the fugitive couple onto the steamship that took them abroad.

"Finding themselves in America, they continued to work as members of the Bolshevik Party . . . Karl Talent took part in founding the Communist Party of the USA. . . . For his activity, he was fired from his work; he and his wife and their child were evicted from their home." At that time, Arthur Talent was a small child. The family lived in the Boston neighborhood of Roxbury.

The year 1923 brought a big change in Arthur's life. "Ill with cancer, Karl Veniaminovich Talent . . . sent his wife . . . with their six-year-old son to the Soviet Union, to Moscow, where her mother, two brothers and two sisters lived." The family were all veteran revolutionaries; one of those brothers had even been chief of Lenin's bodyguard.

"In coming to the USSR, [Elena] Talent had with her a scrap of silk cloth showing her membership in the Communist Party of the USA, which in 1923–24 [was] . . . exchanged for a membership card of the Communist Party of the Soviet Union. Her husband Karl V. Talent died in 1923 in Boston. . . ."

It is easy to imagine why the young Elena Talent and her son Arthur joined the thousands of sympathizers from all over the world who flocked to Russia after the Revolution. As a young widow she had personally experienced the repressions of Tsarist Russia, and of the brutal post–World War I Red scare in the

United States, in which thousands of leftist immigrants like the Talents were arrested and threatened with deportation. Now, leaving the United States for the new Soviet state, she must have felt that she and her son were at last headed for a place of safety, the very capital of the promised land, where well-placed family members and a Party membership card were waiting for her.

Elena Talent and her young son settled in Moscow. Fifteen years later, when the Great Purge was under way, they were both arrested. As a Party member since the year 1907, Elena Talent was clearly destined for trouble. Women were less likely than men to be shot, however; after being tortured in prison, she was sent off for ten years of exile—ironically, to the same part of central Siberia where her husband had been exiled under the Tsar nearly thirty years earlier.

"In the spring of 1948, my sister was freed and came back to me in Riga," writes Minna Gabalina. But in the last few years of his life, Stalin ordered new waves of arrests. Often the secret police filled their quotas by simply rearresting people who had been arrested in the thirties—"repeaters," they were called when taken in for the second time. If they were spies then, the reasoning went, they must be spies now.

"In January 1949," her sister continues, "[Elena] Talent again was imprisoned in Riga Central Prison—the same prison to which I, from 1908 to 1910, still a girl, went with my sister on visits and carried messages to political prisoners who were members of the [Bolshevik] Party." Elena Talent was again sent back to the same part of Siberia. She was not released until five years later, after Stalin's death. "After sixteen years of suffering, [Elena] Talent came back to Riga in 1954, completely ill and not of sound mind, and died on March 9, 1955. She found out nothing of the fate of her son."

12
Library
of Death

It is in such things as the yellowed sheets of paper in Arthur Talent's file, covered with an NKVD man's scrawling handwriting, that we can see Stalinism caught naked. For we are not looking at a public spectacle, a show trial, a presentation to an audience. We are looking at something specifically intended *not* to be shown to the public: the victim was shot; his dialogue with his executioners was locked in top-secret files for half a century. We are behind high walls, inside the machinery of mass murder. And what is it, exactly, that we see? Horror and cruelty, to be sure. But, beyond the terror, there is a note of mystery.

Why was Arthur Talent so persecuted when he was in no way a threat to the state, to Stalin's power, to Soviet communism? Why was so much police manpower spent in extracting his nonsensical confessions? Why was the confession then kept secret?

Consider: here are officers up to the rank of major, trying to get a twenty-one-year-old artist and former elevator operator to confess to being a spy for Britain and for Latvia—countries he had almost certainly never set foot in. And to confess that his spy ring was based in a theater. And to confess that these actors from a tiny country were conspiring to set up a Greater Latvia on Soviet territory. At most times and places, a rational adult would not believe such things.

In its scope and methods, the Great Purge was like the Inquisition. However, the similarity is not complete. It is a pity to dignify the Inquisition by saying that it had any element of rationality, but at least some of its victims were dissenters from the religious orthodoxy of the day. This was not true in the Purge. By the time it began, all open heretics within the Soviet Union were in jail or dead. Almost all the victims were dutiful citizens, and many, like Arthur Talent's mother, were ardent believers. Looked at closely, the accusations against Arthur Talent—of being a spy for a great power (Britain), and of somehow conspiring to convert a small state into a large one (Greater Latvia)—have less to do with heresy than with magic.

The closest parallel to the Purge is in the great witch craze of early modern Europe. There, too, the victims were accused of possessing magical powers. And there, too, the whole epidemic is still difficult to understand. Today, more than two hundred years after the last European witch was burned at the stake, there is still no consensus among historians about the causes of the frenzy, which took at least sixty thousand lives. After tentatively suggesting possible explanations for the beginning of the hysteria, even Hugh Trevor-Roper, the most thoughtful contemporary scholar of the witch craze, frankly confesses that he finds it "mysterious" why it faded out in the 1700s but not earlier.

Part of the problem of explaining mass hysteria is that it has momentum: any outbreak seems quickly to become independent of the causes that triggered it. The hysteria touches an inflammable part of the human psyche, which, once ignited, is hard to put out. Belief in a devil can be as attractive as belief in a god. Even in the best of times, we have plenty of nameless frustrations and fears it is useful to have someone to blame for. And so mass hysteria takes on a seductive life of its own once a class of scapegoats for all problems is officially designated: Witches! Enemies of the people! Off with their heads! The contagion then often lasts long after the specific fears that caused it have disappeared or been replaced by others.

We tend casually to call any kind of search for scapegoats, such as McCarthyism, a witch hunt, but the resemblance between the Great Purge and the European witch hysteria goes far beyond this. Both movements depend on villains who are infinitely numerous and *hard to identify*. In this, both the Purge and the witch craze differ from persecution unleashed against easily recognizable villains, such as Jews, blacks, Armenians, or people who take the Fifth Amendment. Witches, writes Trevor-Roper, "were found to be everywhere, even in judges' seats, in university chairs and on royal thrones. But did the campaign against the witches in fact reduce their number? Not at all. The more fiercely they were persecuted, the more numerous they seemed to become."

What kind of society spawns such hysteria? The Europe of the sixteenth and seventeenth centuries suffered a series of cultural shocks: the Thirty Years' War, the Plague, economic depression, the Reformation and Counter-Reformation. But one thing which remained constant throughout was a cosmology in which all disaster was the conscious, deliberate work of the Devil. The Devil's agents were witches, who caroused with him at a midnight Witches' Sabbath, where he gave instructions on such matters as how to transform humans into toads. An advantage of having enemies endowed with magical powers is that you can accuse them—unlike more traditional and earthbound villains like the Jews, say—of causing absolutely anything. Witches were blamed for attacks of epilepsy, for great storms, for cows that died unaccountably, for capsized ships, for bridegrooms impotent on their wedding nights. And so, too, the agents for the Devil of the Purge years (foreign capitalist powers) were charged with derailing trains, stopping assembly lines, putting nails in the butter, hijacking supplies of bread, and causing all the many other problems of the Soviet economy in a convulsive decade.

With witches, as with Purge victims, concrete evidence was hard to find. The authorities could not produce flying broomsticks in court any more easily than the codebooks or weapons that would prove an NKVD prisoner to be a foreign spy. But in both epidemics, that just increased the importance of denunciations. Encouraging denunciations helped draw millions of additional people into the climate of paranoia. And since denunciations were a sign of good citizenship and could come from anyone, even the humblest peasant could help defend society from witches or from capitalist agents.

Once denounced and seized, both witches and Purge victims were forced to confess. In the topsy-turvy moral universe of hysteria, where the grievances are hazy and the villains hard to identify, a confession is proof that you've caught the right person. More important, it is proof that he or she is the guilty one, not *you,* for getting swept up in the hysteria. Extracting a confession is the uneasy persecutor's form of denial.

To produce those confessions, both witch hunts and the Purge depended heavily on torture. And despite the gap of centuries between them, both epidemics mainly relied on exactly the same technique: using relays of interrogators and depriving the victim of sleep. In sixteenth-century Europe, it was called the *tormentum insomniae.* In the interrogation rooms of the NKVD four hundred years later, it was called the "conveyor belt." After being kept awake for several days and nights, almost anybody will confess to anything, whether to flying through the sky on a trident or to being a Latvian spy.

Also, of course, people confessed because others had confessed, because it was what their accusers wanted, and because they hoped it might save their lives. Accused witches who vigorously denied everything were more likely to be killed. The same was true in the Purge. "The absolutely certain way for a defendant to get himself shot," observes Conquest, "was to refuse to plead guilty."

One final similarity between the Great Purge and the witch hysteria is that the slightest expression of doubt in the overall enterprise was absolute proof that you were guilty yourself. The only top Soviet official who ever questioned the correctness of the Purge trial verdicts in a public speech—Grigory Kaminsky, the country's minister of public health—was arrested the very same day and was never seen again. Judges thought too lenient on witches were often burned at the stake themselves. On the front page of the *Malleus Maleficarum,* the first printed encyclopedia of demonology, was the epigraph: "To disbelieve in witchcraft is the greatest of heresies."

Back in the museum of modern witchcraft, the Moscow office of the KGB, more files lay on the table before us. The next case in the pile, like that of Arthur Talent, also involved a young, American-born man and his mother. But this mother and son were to have one last, horrendous encounter—in prison.

Like Arthur Talent's story, this one also concerns parents who fled from the old Tsarist empire to the United States, had children there, and returned to Russia after the Tsar's overthrow. Victor Tishkevich-Voskov, the subject of the file now open, was born in New York in 1913. Four years later, his parents took him and his infant sister back to Russia on the eve of the Revolution.

The Revolution was soon followed by the Russian Civil War, several years of brutal fighting between the Reds and the Whites—a Western-backed coalition of Tsarist loyalists, ethnic nationalists, and semi-independent warlords. The war claimed millions of casualties, among whom was young Victor's father, who died of typhus in 1920, as commissar of a Red Army division. Victor's mother, Stanislava Tishkevich-Voskova, was also an active revolutionary, the file reveals, traveling on missions to Lithuania, Romania, and Turkey during the Bolsheviks' attempts to organize sympathizers in surrounding countries. Later, she lived at one of Moscow's most exclusive addresses, the Metropol Hotel. This elegant relic of pre-revolutionary days was home to many foreign Communists and senior Soviet officials. Stanislava Tishkevich-Voskova probably earned her room there because she was an old revolutionary and the widow of a Civil War hero.

In 1938, her son Victor was twenty-five years old, living in an apartment on Gorky Street with his wife and their four-month-old son. He was a student at the Moscow Aviation Institute. It must have taken pull to get admitted there, for aviation was enormously prestigious in the Soviet Union then, much as space exploration has been in recent years.

As the year 1938 began, widening circles of arrests swept through the country. Victor Tishkevich-Voskov surely felt some sense of dread. He was in danger three times over: One major category of victims was the foreign-born; another was the families of old revolutionaries like his mother; another, peculiarly, was people connected with aviation. The famous aircraft designer Andrei Tupolev, for instance, was seized and charged with selling secrets to the Germans. (On the advice of a cellmate, he confessed—and lived. He continued designing aircraft, in a special prison.)

Victor was studying full-steam: besides being at the Aviation Institute, he was also taking courses at the Physics and Mathematics Faculty of Moscow State University—another much-sought-after place of study. "And in the last months before his arrest," according to a letter from his widow years later, asking for his rehabilitation, "he was working on building a dirigible of his own invention. They were interested in his project ... and a day was appointed for his presentation before a group of professors. . . . The eve of that day, he was arrested."

Victor's mother, Stanislava Tishkevich-Voskova, had been arrested a week earlier. Apparently her son made no change in his plans to present his design for the dirigible. Perhaps he felt that by excelling in that, he could save his mother. There is no way of telling.

The official record of Victor's arrest shows one of the witnesses to be his apartment building's janitor, an occupation where one was often conscripted for this role—in one of her poems, Anna Akhmatova refers to "the janitor's fear-whitened face." "It could never be said," writes Nadezhda Mandelstam, "that anyone had just disappeared at the dead of night without benefit of warrant or witnesses. This is the tribute we pay to the legal concepts of a bygone age."

Four days after Victor Tishkevich-Voskov's arrest, an NKVD sergeant interrogates him:

QUESTION: You have been arrested for espionage activities in the USSR for a foreign power. We request true testimony.

ANSWER: Yes. I plead guilty to being an agent of German intelligence since September 1937 and to spying for Germany.

QUESTION: Who recruited you to spy for Germany?

ANSWER: To spy for Germany I was recruited by my mother.

This raises, once again, the much-argued question of why victims of the Great Purge confessed so readily to absurd crimes. Many, of course, were beaten or kept awake for days on the "conveyor belt." Interrogators also used trickery and threats. The sergeant may have told Victor that his mother had already confessed. In which case he may have figured that he would have a better chance of saving himself if he confessed, too. Or perhaps the sergeant told him his mother had already been shot—and he figured that therefore nothing he said against her could cause her harm. Or the NKVD may have threatened—as they did with many prisoners—to arrest Victor's wife and child if he did not confess, and he felt he had to weigh his mother's life against theirs.

In this same early interrogation session, here is the sergeant's last question and Victor's answer. Victor seems to be hoping for time to figure out the additional admission that would finally satisfy his tormentors.

QUESTION: You are concealing from the inquest the facts of your espionage activities and the espionage activities of people known to you. We demand true testimony from you on this matter.

ANSWER: I conceal nothing from the inquest and tell only the truth. Probably I've omitted some things from this account of my espionage activities, but I will try to recall them and to report them during the next interrogation.

A skeptic might ask: Are these confessional dialogues real? They are clearly not complete, verbatim records; in one file, for example, the record of an interrogation that lasted two hours (the transcripts show starting and finishing times) took me less than five minutes to read aloud. So presumably much is left out, even allowing time for the interrogator to write everything down, and for the frightened prisoner to read and sign his answers. Some survivors say they were forced to sign confessions that NKVD officers had written out beforehand. Could the police have prepared in advance these question-and-answer dialogues with Victor Tishkevich-Voskov and Arthur Talent, all crafted to show the interrogators skillfully closing in on dangerous spies? Possibly, but I think it more likely that these transcripts are condensed versions of actual conversations. They do not sound scripted, because the talk keeps wandering off the track. For example, at this point in Victor's series of interrogations, the police were trying to build a case that he was a German spy. Yet the only foreigners he could manage to name were not Germans, but Americans:

QUESTION: Who visited your apartment?

ANSWER: Our relatives visited our apartment, as well as my mother's foreign friends, in particular Goldwater, an American, who left for the USA in 1931, and the American writer Williams as well.

There is no clue who Goldwater was. Williams, a later reference makes clear, was Albert Rhys Williams, an American journalist who was long one of the Soviets' most dependable sympathizers. For example, in a 526-page book published in 1937, his only visit to a prison or labor camp is to the famous rehabilitation colony at Bolshevo, a Potemkin-village project maintained for showing off to naive visitors from abroad. "The new colony has ... wide forests and fields to roam in, the river Moscow and its grassy slopes for sun and water bathing. No walls or bars, no guards or keepers, and, after the first year, almost uncurbed freedom.... If [the prisoners] like, they can marry ... fellow-inmates or peasant girls from the villages around. And that, in Russia, means babies, kindergartens, schools. In reality the whole colony is a vast school. So attached to it do many become that on 'graduation' they prefer to stay close by, away from the town and temptation." With rosy glasses like these, Williams was an unlikely spy. But he was still a foreigner, and therefore, to Victor's interrogators, suspect. So they noted down his name.

Finally comes the most stark and chilling piece of paper in Victor's file. It is headed: *Record of the confrontation interrogation of Victor Semyonovich Tishkevich-Voskov and Stanislava Yanovna Tishkevich-Voskova.*

QUESTION TO TISHKEVICH-VOSKOV, V.S.: Do you know the citizen sitting before you?

ANSWER: Yes, I do. She is Stanislava Yanovna Tishkevich-Voskova, my mother.

QUESTION TO TISHKEVICH-VOSKOVA, S. YA: Do you know the citizen sitting before you?

ANSWER: Yes, I do. He is Victor Semyonovich Tishkevich-Voskov, my son.

QUESTION TO TISHKEVICH-VOSKOV, V. S.: What do you know about the espionage activities of your mother?

ANSWER: ... By political conviction, my mother is a backer of counterrevolution. She has a hostile attitude to the actions of Soviet power and the [Communist Party] ... as a result she bent me to her will and I turned into a person hostile toward Soviet power. ... My mother declared that she is a German intelligence agent and demanded my help in collecting espionage information.

QUESTION TO TISHKEVICH-VOSKOVA, S. YA.: Do you confirm the testimony of Tishkevich, V.S., your son, that you involved him in espionage activities?

ANSWER: I deny what my son has testified to ... since I did not involve him in espionage activities.

The policemen ask the mother one more time if she is a spy, and again she denies it. Then the interrogation record ends.

"Confrontation interrogations" were one method interrogators used to extract the signed confessions they sought so desperately. In her autobiography, Eugenia Ginzburg describes two "confrontation interrogations" where she was faced with former colleagues of hers on a newspaper staff who accused her of being a counterrevolutionary plotter. One whispered to her, when the interrogator was distracted for a moment, "Forgive me, Genia. We've just had a daughter. I have to stay alive."

Were Victor and his mother able to whisper a word of explanation to each other? Was he able to say, "Forgive me," and she, "I understand"? Probably not. The record of this "confrontation interrogation" shows that not just one interrogator was present, but two, a sergeant and a lieutenant. They must have calculated that the mother would confess to being a German spy when confronted with the son's accusation—or with the marks of torture on his body. Perhaps because it didn't come out as planned, the interrogation was apparently cut short: the transcript runs only a page and a half. After this brief meeting, there is no evidence that mother and son ever saw each other again. Both were shot.

As a veteran revolutionary, who had lived abroad, Victor's mother knew she had two counts against her, although as a woman she stood a good chance of being sent to the *gulag* instead of being shot. But she made her death much more likely by refusing to confess to anything. Scholars and survivors both estimate that only about one in a hundred Purge victims refused to confess. "Die in silence," was the message on a piece of paper furtively slipped to Rubashov, in Koestler's *Darkness at Noon,* in the prison barber shop. Very few had the fortitude to do so.

For more than fifty years, the record of Stanislava Tishkevich-Voskova's bravery has been kept secret. "The Party promises only one thing," the interrogator in Koestler's prophetic novel tells Rubashov. "After the victory, one day when it can do no more harm, the material of the secret archives will be published. Then the world will learn. . . . And then you, and some of your friends of the older generation, will be given the sympathy and pity which are denied to you today."

The cases of people like Arthur Talent and Victor Tishkevich-Voskov are reminders of the special fury the Purge unleashed against anyone with the slightest connection to a foreign country. *Gulag* memoirs by Russians are filled with descriptions of meeting French, Dutch, or German Communists bewildered at finding themselves in a Moscow prison cell or in the snows of Kolyma. And any Russian who had even passing contact with foreigners was in mortal danger. One prison account describes an elderly inmate "who, when asked why he was in prison, always explained that he was the brother of the woman who supplied the German consul's milk."

Russia's ancient xenophobia helped fuel these suspicions, but as the 1930s drew to a close, Stalin had new reasons for his hostility toward foreign Communists. He knew that many of them would be appalled by his 1939 pact with Hitler to divide Eastern Europe between themselves. All this meant a high death rate among foreign leftists living in Russia at the time. Plus, the Russians had no tolerance for any independent thinking in any other country's Communist Party. The Soviets, Conquest points out, "skimmed off a lively section of the European revolutionary Left, which might have otherwise fertilized a broad and unified movement and barred the way to Fascism."

The Soviet paranoia about any kind of foreign connection shows up throughout the case files the KGB gave me. For example, the warrant for Victor Tishkevich-Voskov's arrest notes suspiciously that, besides being wanted as a German spy, "Tishkevich-Voskov has close ties with Italians."

The main Italian he had close ties with was his wife. Ornella Miziano and her parents had come to the Soviet Union as leftist refugees from Mussolini's Italy. In Moscow, she and Victor were fellow students in high school. Many years later, in 1954, Ornella Miziano wrote to the authorities giving these and other details of her life, pleading for her husband's rehabilitation and for information about his fate.

In her appeal, Ornella Miziano tells the authorities that she is a loyal Communist Party member. She says that she particularly wants to clear Victor's name for the sake of their son, who was a baby at the time of his father's arrest.

"My son is now sixteen and is preparing to enroll in an institute. Not wishing ... to traumatize the child, I have hidden from him the fact of his father's arrest, saying instead that he died. Now that he is becoming an adult, I am faced with the prospect of telling him the truth. But that truth, I am convinced, includes the fact that his father, by his convictions and outlook, by his ancestry and upbringing, could not be a traitor and a counterrevolutionary."

A year after this appeal, and seventeen years after their deaths, both Victor Tishkevich-Voskov and his mother were found innocent of being German spies.

But there is one last twist in the story. In the mid-fifties, de-Stalinization was only at its timid, uneasy beginning. Soviet officials were handling the whole business very warily. When they did posthumously rehabilitate someone, they often did not tell the survivors. Unlike the rehabilitation certificates in the other secret police files I saw, Victor Tishkevich-Voskov's is stamped "Secret." Although the authorities cleared his name, they may never have told his widow.

After we had finished looking through these files with the two KGB officers, Kathleen Smith and I made an appointment to come back the next day for our promised look at the archives.

Olga Matlash from the prosecutor's office met us at the KGB building the next morning. As we sat in the waiting room along with a few other visitors, I wondered who they were and what they were coming to see the KGB about. Matlash had said very little in yesterday's meeting. But as we waited today, she chatted with startling frankness about her husband, who had recently quit his job as a Moscow police detective, her deep worries about the economy, and her hope that her son could get his schooling abroad.

After a few minutes, the smiling, black-haired Major Kirillin and the lean, slight Colonel Grashoven appeared. They ushered us back into yesterday's birch-paneled room. Major Kirillin was still in a suit; Colonel Grashoven apologized for being in what was apparently his normal workday attire, a short-sleeved blue shirt. The major gave me photocopies to keep of the files we had looked at the day before, on Arthur Talent, Victor Tishkevich-Voskov, and several other people.

Still eager to impress us with the KGB's new look, the two men seemed happy to talk about anything that might be on our minds, from the rule of law to the KGB's role in a reformed Russia. I mentioned our own Freedom of Information Act. They claimed to know about it, although I'm not sure they did. Like many people once active in the civil rights movement and against the Vietnam war, I have used the Act to get my own FBI and CIA records. When I described this, both KGB officers looked uneasy and said nothing; they were clearly not sure what a polite response should be.

Then it was time for our "excursion," as they called it, to the archives. The colonel led us into a pleasant, leafy courtyard to the side of the building. He pointed out the second-floor balcony from which Napoleon allegedly watched Moscow burn. ("They're rehabilitating his room also," joked Olga Matlash.) From the courtyard, we entered another building. Both buildings were part of a large complex on this block that belonged to the KGB, said the colonel, and both had once contained prison cells in their basements.

The adjoining building had also been a pre-revolutionary mansion. In the main entrance hall, a wide staircase led to an officers' club upstairs. On the walls downstairs was a photo exhibit about the life of Felix Dzerzhinsky, showing heroic scenes from the Russian Civil War of big rallies, primitive armored cars, and men with fur hats, heavy greatcoats, and long, old-fashioned rifles, their chests crossed with bandoleers. There was something unexpectedly sad about seeing all these defiant-looking, fur-hatted men willing to risk their lives to create this regime whose most shameful secrets were now, from this very building, being shown to the outside world.

On one side of the hall where these pictures hung was a metal door that led into the archives. This was, the colonel said, the first time any American had been here.

The door opened into a large room with thick walls of plaster, containing absolutely nothing except metal shelves that reached to the ceiling. There were no pin-ups, no calendars, no charts, no slogans or decorations. Each shelf was filled with identical tan rectangular cardboard boxes, on end, containing files. All the boxes were marked with several sets of numbers and with the Russian letter for P, for *peresmotreno* [reviewed], meaning that the cases in this room had all been officially looked at since Stalin's death, and almost all the defendants rehabilitated.

Major Kirillin and Colonel Grashoven invited us to look at any file we wanted. Not knowing whose file was in which numbered box (where was Raoul Wallenberg's, for instance?), we couldn't search for a specific case. I pulled a file off the shelf at random and began reading. It was the case of one Vasily Skalkin, sentenced to be shot on November 15, 1937, accused of being a Japanese spy.

"Ah," said Major Kirillin, "he was probably a *Harbinyets.*"

I read further; he was right. Like Olga Infland, the former Karaganda prisoner I had talked to, most people who came to Russia in the 1930s from Harbin, China, ended up in the hands of the NKVD. They were usually accused of spying for Japan.

From two sides of this room, doors led to similar rooms, each with shelves of files on executed or imprisoned people reaching to the ceiling. How many lives were recorded in this library of death? The number of boxes alone was clearly in the tens of thousands, and the boxes I looked in each contained several cases. The officers' statement that this particular archive held some 120,000 cases seemed plausible.

I had imagined these archives for so long as somehow a more sinister-looking place, the very heart of the Great Silence. In the end it was just an ordinary room full of cardboard boxes. Two young men were on duty there, evidently with the job of retrieving files requested by KGB officials. Both wore jeans. Nothing would have distinguished them in a crowd of other blue-jeaned young Russians on the Moscow streets.

Looking at the endless shelves of case files, it felt numbing to think of all these lives cut short. I took some pictures, then we walked into the courtyard again. Colonel Grashoven picked a flower off a bush for each of the two women; finally we said goodbye. Coming back out into the sunlit street felt like emerging from an abyss. The world of interrogation rooms and execution cellars suddenly felt far away. The street was full of noisy traffic, shoppers, and sidewalk peddlers selling everything from shoelaces to radishes. I walked a block together with Olga Matlash, and then we headed in different directions, I for home and she to the prosecutor's office.

My mind went back to a conversation I had had with this friendly, most unprosecutorial woman a few days earlier. It had been a short, formal meeting at her office, to arrange these visits to the KGB. I had shown Matlash my credentials as a correspondent, and she had given me some statistics about rehabilitations. At the very end, I asked her the same question I asked almost everyone: Had anyone in her own family suffered under Stalin?

She looked surprised, paused, sighed, and finally said yes. "My grandmother. She hid it from us. When she was dying, she stayed at my place. She knew my job dealt with this issue. And she told me that she, too, had spent time in prison."

Her grandmother had never told this to anyone else, Matlash said; the arrest had come when she was a very young woman. Released from the *gulag,* her grandmother married, but she concealed her imprisonment from her children. When these children grew up, Matlash said, they "had good jobs" that might have been threatened if it were known that their mother was a former prisoner.

Matlash thinks her grandmother may have suffered because of her class origins: her father was a pre-revolutionary shop owner. The family was Greek, and Matlash does not know how the secret police spelled the name in Russian in their files.

"I am looking for the case," she said, "but so far I can't find it."

13
An
Empty
Field

With the last of winter gone from Moscow, paradoxically the city seemed more gloomy. The snow and ice had covered up potholes, cracked sidewalks, and peeling plaster facades. Without its protective white covering, the capital now looked battered, as if a giant hailstorm had passed through. One day a notice appeared in the hallway of our apartment house, telling us "Comrade Tenants" that the hot water would be shut off for the next three weeks. At first I thought this was some quirk of our building or neighborhood, but it seems that this happens every year throughout the city. The pipes in Moscow's centralized hot-water system are so corroded they have to be cleaned out annually, before the valves clog. And so we heated pots and pans of cold water on the stove each night, and took our three-inch baths.

But the city's gloom ran deeper. The increasing chances to open long-secret files and other doors onto the past was one of the few parts of Russian life touched by any sense of progress. The present, by contrast, as I paused to take a final look at the city where we had lived for nearly five months now, seemed ever more depressing. As spring came, the intractable economic problems loomed steadily larger. The novelty of *glasnost* had long since worn off. With every month that passed, the lines at grocery stores were longer, the clerks more grumpy, and there was even less food behind the counter. On the rare occasions I saw whole chickens on sale in the state food shops, they looked skeletal and scrawny, as if they had come from some *gulag* for poultry. A magazine editor invited me one day to a weekly editorial meeting, where the staff goes over headlines and layouts. When they came to one group of pages, the editor in charge of the section wasn't there. "He's across the street buying sausages," someone explained. Nobody was surprised: if you hear of a place where meat is on sale, that takes priority over anything else. When we were invited to people's homes for dinner, they often apologized that there was no meat. If there was, it was admired and celebrated, with jokes and stories about how much bartering and favor-trading was necessary to get it.

It was striking how, even about smaller things in a state of collapse, nobody did anything. American pundits preach that the Russians need a sense of private property. But they miss the point. Russians have a very strong sense of private property. I hardly ever visited an apartment that wasn't immaculately kept, with one glass-doored bookcase full of the Russian classics and another displaying knickknacks and figurines; people usually had special slippers, called *tapochki,* ready for visitors, so you could leave your shoes at the door and not track in dirt from the street. It was their feeling about public property that was the problem. Until this changes, real democracy in Russia, a democracy that does not just go through the motions of holding elections but is built upon people's confidence that they themselves control the state, and not somebody else, is still far off.

The apartment house where we lived, for example, was at Moscow's equivalent of a Park Avenue address: near the city center, on the Moscow River, right across the street from what was then Boris Yeltsin's office. But the building's public areas were an unbelievable mess. To enter the narrow front hall of our entry, you had to dodge the open, swinging metal door of an electrical junction box full of exposed wiring. The stairway was covered with cigarette butts, empty bottles, apple cores, and an old, rusted radiator lying on its side. Where the stairs reached our floor, there was a hole in the concrete big enough to drop a baseball through to the flight below. The elevator smelled of garbage and worse, and looked as if it hadn't been cleaned since World War II. On one side of the building was a courtyard strewn with piles of debris; on the other a dirt depression filled with pools of stagnant rainwater. These bred mosquitoes, who found their way into our bedroom, to lie in wait to attack after dark. Theoretically the building had a janitor, but we never saw him.

Beside our entry door, day and night, big rats scurried in and out of a hole to the basement. At one point they gnawed through a telephone cable, knocking out phones for ten days. Why didn't anyone protest the mess? Why not put out rat poison, drain the mosquito pools, have a cleanup day, take turns sweeping the stairs, rouse the janitor?

We were talking about this one day with some Russian friends who came over for lunch. If things were this bad in the United States, I said, people might start a tenants' union. One technique these have used is to keep putting dead rats on a landlord's doorstep until he does something.

One guest tried to explain why that could never happen here. "You'd take your dead rat to the District Executive Committee. They'd tell you, 'We've asked the Sanitary Station to send in a team to kill the rats.' Then the Sanitary Station would tell you, 'We can't get the authorities to give us any rat poison.' Then the rat poison factory would tell you, 'We can't get the foreign exchange to buy the right chemicals, so now we're making fertilizer.' Then you'd go to the prime minister, and he'd tell you, 'Rats? I've got worse problems!' And you'd take your dead rat back home. . . ."

The harsh despotism of forty years ago seems a long way from the bedraggled Russia of today, where every bureaucrat claims, "It's not my department." But, of course, decades of the one paved the way for decades of the other. In the long run, history catches up with you. When all power flows down from the top, no one at the bottom—or even in the middle—dares take any initiative. And now that long-time Communist Party rule has collapsed, the passivity of every other level of society is laid bare.

It is not only the Party itself that has evaporated, but also the last shreds of the dream it promised, which inspired so many people in the early part of the century. For Russians, the Soviet experience has killed, besides the 20 million dead, the idea of Utopia—or at least of one that could be built on earth, by constructing a more just and fair human society.

The gaping vacuum this has made has left many Russians turning to the supernatural. From time to time, chain letters appeared in our mailbox, each promising health, happiness, and good fortune if only we didn't break the chain. One letter claimed that Khrushchev had broken just such a chain—and two days later was ousted from power. I'm sure you could chart a correlation between the steady drop in the economy and the steady rise in the number of gurus offering salvation or magical rewards. Hare Krishnas appeared on the streets of Moscow along with the spring blossoms; seers and hypnotists performed on TV. Werner Erhard swept into town from California to begin a big Moscow operation.

If salvation didn't lie in the teachings of some huckster from abroad, it lay half-mystically in the pre-Communist past. Daniil Granin's biographical novel *The Bison,* an admiring portrait of a scientist persecuted by Stalin, makes much of how the great man's fearless spirit is built upon his ties to his forebears, landowners and Cossack commanders. "His great-great-grandfathers were right behind him, not dusty ancestors, but living relatives . . . they all made up his past, his roots in this world, and their blood flowed through his veins, their genes lived in him, he was their continuation." At a right-wing demonstration near Red Square, we saw several young men dressed in the uniforms of the Russian army that defeated Napoleon.

Dramas about no less than six different Tsars, from Ivan the Terrible to Nicholas II, played in Moscow theaters in 1991. There is a movement to make Nicholas II, who was murdered along with his family by the Bolsheviks, a saint of the Russian Orthodox Church. Popular ballads now praised the soldiers of the White Army. And, some suggested, they lost the Russian Civil War to the Reds because they fought in too gentlemanly a way. In the pedestrian underpass under Pushkin Square, a vendor was selling framed photographs of the last Tsar and his wife and children. The young princesses, in long white dresses, were playing on a lawn.

The underside of this glorification of Imperial Russia is a nasty racism toward non-Russians. Muscovites would talk angrily about "blacks"—by whom they mean the Armenians, Azerbaijanis, and other ethnic groups, mostly from the Caucasus, who sell food from the trading stalls of the city's open markets. They resent these people for doing well in the new economy, while ordinary Russians are getting poorer. There is even a current expression in Russian which people use when something goes well for a change: "Now we're living like white people."

In another way, as well, Moscow was becoming an *apartheid* city, although the dividing line was not racial. It ran between those who had foreign money and those who did not. As foreigners, we lived comfortably because whenever we wanted we could shop in the special hard currency stores that sold French cheese, Scandinavian fish, German delicatessen meats, and other imported food. But, I noticed, among the other shoppers spending their dollars or marks or francs in these sleek mini-supermarkets, there were more and more Russians. Some were clearly people paid in cash by foreign tourists—hotel doormen, cab drivers, prostitutes. Others, especially those who drove up to the store in Volvos or BMWs, were more likely from a more privileged class—long-time Party members, managers of large enterprises who, as the country practices the privatization so urged on it by the West, are happily privatizing these factories into their own hands, and pocketing some of the hard currency profits. At the funeral of communism, goes one joke, a lot of people are hopping out of the coffin and into the procession. As the old Russia crumbles, the new one replacing it will have a vastly, ominously greater gap between rich and poor.

"When thinking about the fall of any dictatorship," writes

Ryszard Kapuściński about Iran, "one should have no illusions that the whole system comes to an end like a bad dream with that fall.... A dictatorship ... leaves behind itself an empty, sour field on which the tree of thought won't grow quickly. It is not always the best people who emerge from hiding, from the corners and cracks of that farmed-out field, but often those who have proven themselves strongest...."

May came, and with it the annual VE-Day celebration of victory over Germany. But the holiday no longer provided the emotional boost of old, despite the traditional red flags, parades, and fireworks. For one thing, the defeated enemy was the very country now sending Russia the most emergency aid: hundreds of thousands of parcels of packaged food with brightly colored labels in German. The parcels went largely to old people living on pensions—just the group that includes World War II veterans. But beyond this there was a deeper problem: Russia's new openness about its past meant facing painful truths, not only about the *gulag* and the Purge, but also about World War II.

Since 1945, the cult of the war was at least as important as the cult of Lenin. Honoring the great sacrifices of the war used to be one of the few things that government and people agreed on. Even dissidents seldom questioned the mythologizing, for they, too, had lost fathers and brothers at the front, had seen city after city in the Soviet Union destroyed, and had rejoiced to see the Nazi invaders defeated. "Compared to everything in the thirties," says Dudorov, a former *gulag* prisoner, in Boris Pasternak's *Doctor Zhivago,* "... the war came as a breath of fresh air, a purifying storm, a breath of deliverance."

For more than four decades after the war's end, Soviet citizens celebrated the victory with tremendous fervor. The war was remembered endlessly in novels, films, stamps, and eternal flames. In Moscow and other cities, on any weekend you could see a stream of newlyweds, the brides in white dresses, leaving bouquets of flowers at the local war memorial.

But now, as the fiftieth anniversary of the German invasion approached, hard truths about the war could be denied no more and the press was filled with searching questions. Why was so much of the front-line air force wiped out on the ground on the war's first day? Why were millions of soldiers surrounded and captured in the first months? Above all, if the Soviet Union started the war with thirty more divisions than Germany, more than twice as many tanks, and nearly three times as many warplanes, why did it suffer the deaths of what the government now admits was 27 million people, while the Germans, battling not only Russia but the United States and England as well, lost less than 5 million?

The key to Russia's huge number of casualties, especially the stupendous losses of men and territory at the beginning, is, of course, Stalin's sweeping purge of the Red Army on the very eve of the war. Some 43,000 army and navy officers were shot or sent to the *gulag*. More Red Army *officers* were executed or died in Soviet labor camps than died in the war itself. Among those of higher rank, the figures are even more stark. Some six hundred Soviet generals were killed in action during the four years of World War II; one thousand were shot by Stalin just before the war. Has any aggressor in history ever had so much help from the country it was invading? "What's an army without officers?" asks a retired general in the new Russian documentary film *The Trial*. "Just a lot of mouths to feed." When interrogation records of a few of these officers were recently released by the KGB, said an article in *Pravda*, they had "gray-brown blotches" on their pages. Laboratory tests showed these to be blood.

In the late thirties, German intelligence eagerly watched the purge of the Red Army, and tried to fuel Stalin's paranoia still further by planting false papers that showed various generals to be in touch with Berlin. Meanwhile, NKVD agents, eager to advance their careers by proving that top Soviet generals were Nazi spies, were themselves forging similar documents in German. Today, researchers are still trying to figure out which papers are which. The one thing that just about everybody agrees on is that before the war no top Red Army officers did conspire with the Germans.

The war is fertile territory for revisionist Russian historians. And for psychologists. Why did Stalin ignore all alarm signals of the impending German attack? He failed to put Soviet forces on alert; he dismissed desperate, repeated warnings from front-line commanders and diplomats; when spies brought messages about German preparations, he ordered them shot. The millions of Russians who dutifully condemned Purge victims as foreign spies ignored the danger from within: the madness of Stalin. The dictator himself killed millions of imaginary internal enemies but ignored the danger from without: the 3 million German troops massing on his border. In the end, he was the biggest denier of all.

The most depressing thing in this sad Moscow spring was the number of people who wanted to leave. Among young professionals who spoke a Western European language, it was hard to find anyone who had not thought about emigrating. Hundreds of people stood in long lines outside the U.S. Embassy each day, trying to get visas. And no longer was it just Jews who wanted to go.

So many of those who hoped to leave were the country's best minds. One evening I was invited to a meeting of a monthly seminar at the Filmmakers' Union. The topic was patriotism and nationalism, and the subtle and important differences between the two. People took the discussion very seriously. The fifty or so participants were all filmmakers, novelists, and scholars, many of them well known. They talked for four hours, around long tables arranged in a hollow square, as if for an international peace conference. They spoke carefully, thoughtfully, elaborating ideas for five or ten minutes at a time rather than engaging in repartee. It was free speech at its best, plunging deep into the historical and philosophical roots of present-day problems. One would be hard put to find an equivalent group in the United States.

At one point in the discussion, I made a short comment. As I finished speaking, one of the filmmakers got up, walked over, and sat down in the seat next to me. Was he struck by my brilliant remarks? Did he want to cast me in his next movie? No—he had suddenly realized that there was a native English speaker in the room. He pulled a letter in English out of his pocket, leaned over, and asked in a whisper if I would correct the spelling and grammar. It was to the U.S. Consulate, asking for an immigrant's visa.

"Pity the country," wrote the Marquis de Custine, "where every foreigner appears as a saviour."

The time was approaching for what would mark the end of my stay in Russia, a trip to Kolyma. I gathered names of people to see and made travel arrangements. But this most notorious corner of the *gulag* was at the opposite end of the country, a third of the way around the globe from Moscow; it seemed too bad not to stop at one or two places along the way. I looked at maps, and asked friends for advice. Then one day I heard about a town I knew I had to visit. It was a place where a chilling piece of long-suppressed Soviet history had been exposed with dramatic suddenness, and where the consequences were still reverberating. And it was not a historian who had drawn aside this veil across the past, not an investigative journalist, not a survivor. It was a river.

SIBERIA II

Soviet secret police execution cell, Kiev, 1920.

14
Secrets
of the
Riverbank

Although the network of the *gulag* was spread across every part of the Soviet Union, the major islands of its archipelago were usually defined by a natural resource. One such cluster of labor camps was in the coal-mining area around Karaganda, where I had been. The largest and most deadly group of camps was in Kolyma, where I was going, where prisoners were sent to mine gold. Yet another favorite spot for sending prisoners and exiles—under the Tsars as well as Stalin—was along the upper reaches of the Ob River, in western Siberia. There, most people were put to work cutting timber.

The Ob is one of the longest rivers in Asia. From its headwaters on the Mongolian border, it flows northwest across Siberia, sweeping past mountain slopes and summer dachas, past factories and cities, and then past a long stretch of dense fir forest before it finally empties into the Arctic Ocean. On one of the Ob's tributaries, the Archpriest Avvakum, the great religious heretic of the seventeenth century, was exiled for a time before he was burned at the stake. Stalin himself was banished to a village in this region during one of his terms of exile, in 1912. Also in the territory of the upper Ob was the orphanage where Vladimir Glebov, Lev Kamenev's son, learned self-defense from his Gypsy friend.

One of the principal towns on the Ob is Kolpashevo, which lies some 600 miles south of the Arctic Circle. In the 1930s, among the exiles sent there was Nikolai Klyuev, a distinguished poet of peasant origin, who was later shot. Klyuev wrote to a friend:

I have been sent . . . to the village of Kolpashevo, to a certain and tortured death. Four months of prison and transit-stages . . . have gnawed me to the bone. Remember me in this hour, an unfortunate, homeless old man of a poet. The sky is in rags, the rain is slanting and flies in from thousands of miles of marshes. The wind is never still and this is what they call summer here. Then follows the minus fifty degree winter, and I am naked. I don't even have a hat, and my trousers don't belong to me, because the criminals took all I had in the shared cell. Try to think how to help my Muse, whose prophetic eyes have been viciously gouged out! Where can I go? What can I do? I'm muttering to you, like I do to my heart.

In a searing, unpublished cycle of poems seized when he was arrested, and only recently discovered in the KGB archives, Klyuev almost seems to have had some foreboding of an event that would take place in the riverside town of his exile. He speaks of ash trees "waist deep in blood," of taking a pen "dipped in lava" to record Russia's fate. He writes of Stalin as

. . . the devil lumberjack
[who will] dent the axe on the wild forest of bones
and the impenetrable thickets of graves
that could not be counted.

Finally, almost uncannily, Klyuev speaks of the great rivers of Russia moaning and weeping at what they see along their banks.

Like the Mississippi, which also traverses flat, low-lying country, the Ob flows in lazy bends and loops, overflowing its normal course when it floods. In May 1979, swollen with the water of melting snow and carrying big, swirling chunks of ice that had broken up in the spring thaw, the Ob began eating away at its banks at Kolpashevo. The town sits on some low, sandy bluffs overlooking the river. Kolpashevo's townspeople were used to seeing the spring floods at work, but this time they watched with mounting horror. As the river gnawed away at the base of the bluffs, the earth and sand that crumbled down into the water gradually disclosed a mass of human skeletons. Below these bones was another layer, more ghastly still: hundreds of whole corpses. Embedded in the riverbank were more than a thousand skeletons or bodies in all.

As soon as they heard the news, old people in Kolpashevo came to the spot carrying religious icons. Everyone in town knew why the dead were there, for these bodies were beneath the site of what had been, in the late 1930s, the regional NKVD prison. In the preserved corpses of 1979, some Kolpashevo residents recognized bodies of people they knew, still wearing the same clothes they had been arrested in some forty years earlier. One older woman reportedly identified her husband's body.

Although some of the other mass graves now turning up throughout Russia and the other republics are larger, the one at Kolpashevo has been the most extraordinary of all these grisly discoveries. One reason was that the lowest layer of bodies was so well preserved. These corpses lay at the bottom of the grave, whose floor was the great sheet of permafrost that underlies most of Siberia. In addition, the soil dumped on top of them was dry, sandy, and laced with lime as a disinfectant. The bodies had not decomposed; half-frozen, they had been mummified.

The other notable thing about the Kolpashevo grave is that the government immediately and frantically covered it up again. In 1979, it was six years before the first stirrings of *glasnost*. No open discussion of the Stalin era was allowed. When the bodies appeared in the riverbank, alarmed, high-ranking Communist Party officials flew into town. The KGB cordoned off the site. Special crews of soldiers and laborers worked in relays for some two weeks to destroy all traces of the grave—no easy task when the thousand-odd skeletons and bodies were lodged in the bank of a large, fast-flowing river in flood. The authorities at first vaguely claimed that these might be cattle bones. Finally, fooling nobody, they announced that these were the bones of World War II deserters.

As I collected scraps of information about Kolpashevo, I was particularly haunted by the way that the people buried there had been, in a sense, killed twice: once by the executioners of the Great Purge, and then again by the modern-day KGB, who had tried to sink their bodies in the river some forty years later. For people who lived in Kolpashevo now, coming to terms with Stalinism meant dealing, psychologically, not with stories of far-off concentration camps, but with a mass of corpses beneath the very streets of their town. How did that affect the whole issue of denial and of seeing? Had any of the townspeople been accomplices to this massacre in their midst? Or did they feel like victims too? It seemed a case where the line between victims and executioners might not be so clear.

In a thinly populated part of the country well north of the Trans-Siberian Railroad line, Kolpashevo lies far from the usual travel routes. The jumping-off point for reaching the town is Tomsk, capital of the *oblast,* or province, in which Kolpashevo lies. The flight from Moscow to Tomsk took many hours, much of it over the Siberian *taiga,* the largest forest on earth. The *taiga* is home to half the world's conifer trees, some of them three hundred years old, and is now—to the dismay of environmentalists worried about erosion and global warming—being eyed hungrily by U.S., Korean, and Japanese lumber companies. It was long after nightfall when my plane at last began the descent toward Tomsk. A forest fire was burning far below, a flickering orange patch several miles long, surrounded by an endless ocean of darkness.

Instead of finding the drab gray of most Siberian cities, it was a delight to wake in the morning and discover Tomsk to be quite beautiful. An old university town and river port, Tomsk dates from the 1600s. Much of the city looks as it did more than one hundred years ago: the old governor's mansion, where the Tsar once stayed; a home for invalids from the Napoleonic Wars; and an ancient, glass-roofed botanical garden with what must be the only banana trees in Siberia. Most spectacular of all, Tomsk's older buildings are almost all of wood painted in bright colors, their roofs sprouting a Victorian forest of sprightly turrets, spires, cupolas, cornices, and balconies; their bay windows fringed with festive wood carving that imitated wrought iron or folk embroidery, with hearts, flowers, and animal figures worked into the design.

Over time, the weight and heat of each building thaws the permafrost below and the structure steadily sinks by a fraction of an inch a year. Because most of Tomsk's wooden houses are more than a century old, they have often sunk so far that the ground-floor windowsills are below sidewalk level. The sinking is uneven; the buildings tilt slightly. This gives the place a strange fairy-tale look, as if these houses were a crowd of cheerful dwarfs, dressed in colorful clothes, standing on stubby, unsteady legs and leaning against each other.

At one Tomsk restaurant, there was bear meat on the menu. (Officially the place served no alcohol, but, I was told, if you made the proper arrangements with the waiter beforehand, the samovar would pour vodka instead of tea.) Tomsk seemed all the more magical because the city had been completely closed to foreigners until six months before. After five months mostly in Moscow, it was refreshing to be where few people had ever seen a live American or Western European. Tomsk's cab drivers, unlike the capital's, did not look like Mafiosi, and no black-market moneychangers or hard currency prostitutes thronged the streets. No doubt the package tours to Tomsk will start one day, but in the meantime, without another foreign visitor in sight, the city felt strangely innocent.

Several people in Tomsk had something to say about the mass grave at nearby Kolpashevo. One was Gennady Sapozhnikov, a sixty-one-year-old machinist who lived in one of Tomsk's old wooden buildings. He was a short man with a large nose and whimsical, cheerful eyes. Sapozhnikov was a boy in a village several hundred miles north of here when his father, a farmworker, was arrested in July 1937. "I was seven at the time. They charged him with being part of some kind of monarchist group, wanting to put the Tsar back on the throne. But all this was garbage! An illiterate man, working from dawn to dusk, in the far North—what conspiracy could he possibly be involved in?"

Sapozhnikov's family never knew for certain what happened to his father. But they long suspected that he met his death at Kolpashevo. It was the regional secret police headquarters, and a relative had seen Sapozhnikov's father and other prisoners being taken toward Kolpashevo by river barge. No one ever saw him again.

Then and now, the Ob River and its tributaries are the major north-south transportation route of western Siberia. By coincidence, during the very week in May 1979 when the flooding Ob exposed the grave at Kolpashevo, Gennady Sapozhnikov and his mother happened to be traveling past the town on a passenger boat. It was late evening, but Kolpashevo is far north and the sun was still up. "The Kolpashevo pier is a little bit before the burial place," Sapozhnikov explained. "The boat pulled in. I could hear noises from the shore. The sun was setting; the entire shore was lit up with sunshine. I saw this hastily made fence, and a lot of people. But the boat was blowing its whistle and I didn't dare go ashore to investigate. Then the boat was on the move [past the grave]; the corpses were well lit by the sun." They could not see the faces of the corpses clearly, but looking out her cabin window at the grave Sapozhnikov's mother said, "My Petya is here."

Sapozhnikov is a prolific amateur painter. Some twenty pictures, one above the other, cover the walls of the two-room Tomsk apartment he shares with his wife. They are all straightforward realist oil paintings, mostly of Siberian nature scenes: animals, streams, campfires at dusk. That evening on the riverboat, he had his drawing pad with him and quickly did some sketches. Later, at home, he made a painting. Since the KGB did not let anyone take photographs, this may be the only graphic record of what the river revealed in 1979.

The painting shows the dirt riverbank torn open by the floodwaters, as seen from the river itself. Everything is tinged golden by the slow-setting northern sun. At the upper layers of the grave, skeletons are lying on top of each other in a crazed jumble. Lower down, near the water level, the skeletons give way to actual corpses. A floating boom of logs tied together with cables cordons off the spot from the rest of the river. Inset into the lower-right-hand corner of the painting is a second image: a close-up of several skulls lying together on the steep bank, a bullet hole in each.

Like all the country's major cities, Tomsk has a chapter of Memorial; its co-president was a scientist in his late fifties, named Wilhelm Fast. A tall man with a majestic gray beard, extremely gaunt from a recent illness, he looked like an Old Testament prophet who had just survived some faith-testing ordeal. He is, in fact, a devout Christian. Before we sat down to lunch at his apartment, he and his wife and children bowed their heads in prayer. Fast then sprinkled something that looked like chopped-up clover on top of the soup. It was *cheremsha,* he said, a type of broad-leafed garlic that grows in the wild and is rich in nutrients and vitamins. "It saved many of us," Fast said.

And so, before we began talking about Kolpashevo, I asked Fast about his own story. His family's crime was to be of Dutch and German descent—even though they had lived in Russia for five generations. More than half a million "Volga Germans," as they were called, were one of several entire ethnic groups deported en masse from their homes during the 1940s. "My father and grandfather were arrested on the same day," Fast said. "My grandfather died in a camp. My father was locked up for ten years. I was two years old when they took him away. That picture is recorded in my mind like a photograph. In black-and-white, with shadows. I see them go off. A corridor, with black silhouettes going away. Bright sunshine outside. Darkness inside. My sister was eleven days old. My mother had had a difficult childbirth and was still in bed."

Later the police came for the rest of the family. "We were deported in cattle cars to a village seventy kilometers from Karaganda. We were on the journey for a month. Some people died of starvation on the way. We arrived at Christmas, 1941. They put us in some deserted huts. Not a tree anywhere. There were blizzards. My uncle was not in his right mind: he couldn't work, he just dreamed of food. The men were immediately taken away, to work in the coal mines. My mother went to see the chairwoman of the collective farm. 'There's no wood to heat the hut with,' she said. 'Heat it with snow!' the chairwoman said.

"My mother went to the barnyard at night. All believers know that you mustn't steal. She cried out, 'God, please forgive me!' She grabbed some straw and took it home. There was a little stove; she burned the straw, and we three kids got into a kind of cauldron on top to keep warm.

"All the time, we saw people dying. Then suddenly all women from fifteen to sixty were conscripted to become war workers. They took my mother away. It was the autumn of 1942. I screamed and wept. They just put the women in horsecarts and took them away. My grandmother was sick in bed; she died in 1944. My aunt was a little older than me and she looked after me. But she had TB; she died in 1945. My uncle died of hunger. It was winter—we had to dig up an old grave because it was impossible to make a new one. We found a box to cover his face so the earth didn't fall on it. I began to earn my living when I was eight years old."

Somehow—he credits his religious faith—Fast survived all this to finish school and his higher education as well. He moved to Tomsk and became a professor at the university there, a specialist in the study of meteorite impact craters. But as he examined and measured traces of these over the years, traveling through the Siberian countryside, he quietly gathered information about another kind of impact crater as well. He poked through the ruins of old concentration camps and grave sites, and talked to villagers who had lived through famine, prison, and exile, just as his own family had. Fast heard about the mass grave at Kolpashevo as soon as the river opened it up. Long before *glasnost,* he visited the town, and wrote about it in one of the hand-typed, illegal *samizdat* publications that circulated even here, in Siberia.

"They invited me to emigrate to Germany," Fast said reflectively. "I could have had a nice house and garden. But then who'd do this work? Without us digging underground like moles in the horrible past, up there on the surface flowers won't bloom...."

In 1982, a new round of troubles in Fast's life began. Because of his Christian beliefs, he was forced out of his university post. He found work as a beekeeper, then as a janitor in the apartment building where he lives. Other professors who live there chatted with him as he swept the courtyard. But in the tumultuous last few years of changes in Russia, he has won his university job back and been elected a deputy in the *oblast* legislature. He spends his spare time working with Memorial. When I told Fast I was going to Kolpashevo, he offered to come with me, and I accepted immediately.

15

Two Fathers,
Two Daughters

Wilhelm Fast and I flew from Tomsk to Kolpashevo early one morning on a small turboprop. As a legislative deputy, Fast flies free anywhere in the *oblast.* On the trip north, the Ob River came into view beneath us, a wide band of shimmering water snaking its leisurely way through the endless, surrounding evergreen *taiga,* often meandering off into side channels. Local airplane flights in Russia are like bus rides in Mexico. On this plane, two passengers were carrying small dogs in briefcases, and Fast had a satchel containing a large cabbage. His wife's mother lives in Kolpashevo, and whenever you visit friends or relatives in Russia these days, you bring food. "We've got an apartment full of books on astronomy, meteor craters, mathematics—and we're worrying about vegetables," he said wryly. Tens of millions of Russians now till small plots on the outskirts of all the big cities. Over the weekend, Fast and his wife, a scientist who studies tropospheric clouds, had been out planting potatoes.

Some 37,000 people live in Kolpashevo today. The town is home to various depots and repair shops connected with the timber industry, and to a ski factory that uses some of the wood. We found Kolpashevo in the midst of the short, very green Siberian summer. Some people lived in four-story, rundown, concrete apartment buildings, but much of the town was older: quiet, tree-shaded streets of single-story wooden houses with towering stacks of fire-wood for the winter piled up beside them, their yards surrounded by blue or white picket fences. The streets were mostly unpaved, sometimes with raised wooden sidewalks to keep walkers' feet out of the spring mud.

After dropping off the cabbage with Fast's mother-in-law, we headed to the town's Life-Saving Station, a sort of coast guard out-post on the river. Small freighters, tugboats, and barges passed by; several passenger vessels were docked at a nearby pier. A sailor, shirtless in the summer heat and wearing rubber wading boots, of-fered to take us along the river to the grave site in an outboard mo-tor boat.

The crumbling dirt riverbank at the grave site was perhaps 25 feet high, and I could see clearly how it had eroded over the years. Some wooden houses perched precariously near the edge looked as if they would tumble into the river and float downstream to the Arctic in the next spring flood. The waters had washed away all but a few traces of the NKVD headquarters that once stood here. The sole, ghostly remains were two wooden foundation beams, jut-ting out into space about six feet below the top of the riverbank.

The outboard landed, and Fast, the sailor, and I scrambled up the steep slope. At the top, amid grass strewn with yellow flowers, a skeleton of sheet metal was fastened to a tree. Below it was a metal plaque, with lettering made with a hammer and nail: "In memory of Pyotr Christopherovich Bremerberg. Born Nov. 17, 1888, shot here by the NKVD April 22, 1938, his corpse sunk by the KGB May 9, 1979." I had already seen several new stone mon-uments to Stalin-era dead at various spots across the country, and still more elaborate memorials are under construction. But there was something peculiarly moving about this homemade one. Per-haps one of the dead man's children had been a metalworker.

The sailor from the Life-Saving Station told us about the day the river had revealed the bodies at this spot. "At that time I lived some two hundred meters from here. I came home from a trip, and there were a lot of people right here. They were crying. Probably they had relatives buried here. I went to work and came back in the evening, and saw the place had been closed off, with a fence made of boards.

"A friend of mine lived here, but his house was inside the fence. The soldiers wouldn't let him in. They quickly moved all his things out and gave him a new apartment." He shook his head incredulously. "Because of this business, he got his new apartment."

Inside their board fence, the KGB put bulldozers to work. And for the whole two weeks of the cleanup, the KGB anchored a powerful diesel boat just offshore, with its stern to the bank. They ran the engine, and the wake from the boat's propellers gradually eroded away the bank further, washing skeletons and bodies into the river.

For good measure, the KGB then dumped several bargeloads of gravel on the spot, to cover up any remaining bones. "But the river washed the gravel away," Fast said, "and you could see the bones. Last October, I walked around here and collected a whole bag of bones. The water level is high now. When it goes down, you'll be able to see bones again."

We took the outboard back to the Life-Saving Station. It was a low wooden building with a squawking two-way radio, a cabinet of first-aid supplies, and a stripped-down outboard motor lying on the floor. Waiting for us there was Anatoly Potyekin, who had been the station's chief in 1979. He was a tough-looking man with the tanned, weatherbeaten face of someone who has spent most of his life outdoors. He frowned suspiciously at these two unexpected visitors, a tall, bearded parliamentary deputy and a foreigner. What did we want? Why were we here? But once Fast made clear he was not from the Communist Party, Potyekin shook our hands warmly. Sitting on a tabletop with a navigational chart of the Ob on the wall behind him, Potyekin began talking about the day in 1979 when the river opened the grave. He spoke a rapid-fire, obscenity-laced slang, but Fast rephrased it into a slower, simpler Russian I could understand.

"I was woken up at six in the morning," Potyekin said, "and ordered to come to the KGB. They told us to get anchors and cables ready. We went to the spot. A motorboat was forcing the water toward the bank with its propellers, and the bodies floated past us. An armada of them! It was a nightmare!

"Half the bodies were dressed. I saw them all. And they all had holes in their heads. One or two holes. Fuck them! They'd shot the prisoner in the head, and if he didn't die, then they'd shot him again. All these bodies were sliding past us in the water. They were so well preserved, their mouths wide open—it almost seemed as if they were screaming.

"They washed and washed and washed the bodies out of the riverbank. They told us, 'Your job is to sink them all.' I got angry and said, 'Fuck you! You have no scruples! Why don't we take them out of the river and bury them someplace?' "

But the KGB told him to follow orders. "I sunk them. We were sinking them for two days—any body that was floating, that is, for some went to the bottom immediately. It took us a lot of time. We used wire and metal weights."

Some corpses slipped past Potyekin's crew, however. In the end the KGB's frantic attempt to dispose of all the bodies in the river probably attracted more attention than it would have done to retrieve and bury them. Among other things, the Ob is used by lumberjacks to move newly cut timber out of the forest to a sawmill a few miles downstream from Kolpashevo. Suddenly, among the logs floating into the mill's sluiceway, horrified workers began finding corpses with bullet holes in their skulls.

What Potyekin did after telling his story startled me, although perhaps in this country of poetry lovers it should not have. Still sitting on the table, he leaned back, and in a resonant voice filled with sonorous vowels, he suddenly began reciting a poem he had written after those days:

> I am shot!—lightly clad. They judged me;
> The dull, featureless gun barrels carried out the sentence.
> In the depths, in the quiet, on the bloody sand,
> My last cries were lost.
> I am shot!—alongside other bodies,
> Silent, with open mouths.

Fate, why did you drive us here
Piling one body on another in layers?

Seeing my tape recorder still going, Potyekin launched into another poem, this one about a heroic machine gunner in World War
II. Wait! he said, there are still more. But Fast nudged me; it was
time to go.

One person buried in the Kolpashevo grave, Galina Nikiforova is
certain, was her father.

Nikiforova is a retired Kolpashevo schoolteacher, a pink-
cheeked gray-haired widow in her late sixties. She spoke in a voice
laden with resignation. Her father was a veteran of World War I
and of the Red Army in the Russian Civil War, who was a
Kolpashevo school principal. Late one night in September 1937,
there came what everyone in those days feared: a knock on the
door.

"Father opened it. 'Please come in,' he said. They came into the
dining room and right away said, 'Hands up!'

She raised her hands to show. "I looked at him. My God! At
the front he had never had his hands up. He had such a lost face.
He looked at me with terror. I was fourteen, and I understood everything only too well.

"[The search] lasted till morning. They didn't leave us anything. They took his notebooks. When he'd studied in Leningrad
at a teachers' college, he'd had this black notebook, and we loved to
read it, because he'd listed the operas he'd been to and what singers
sang—Chaliapin and so on. We loved that notebook so much." As
she talked about it all now, more than fifty years later, Nikiforova
wept.

The police also took all the family's photographs of her father,
as if his very image had to be arrested too. Today she has only one
picture of him left, saved by friends. She pointed to it, a somber
young man in his World War I uniform. Sepia-print photos on the
wall showed various other relatives: uncles, in-laws, grandparents.

Many of them, especially the men, had also fallen victim—this man shot, that one sent to forced labor on the White Sea Canal.

On the day the police finally took her father away at dawn, Nikiforova said, "I went to school, and from school to the NKVD. I thought the doors would open and they'd let Father out. I looked through the slits in the fence. To this day I can't stand the odor of musty leaves, because on Lenin Street there, there were a lot of poplars. The smell always reminds me of that sorrow."

After the arrest, word spread that sometimes wives and children were arrested also; Nikiforova's mother kept bundles of clothes packed and ready for them all. Some people shunned them. "We had some very close friends, we were like one family, living in apartments next to each other on the same hallway. The Shipunovs. They stopped saying hello to us. They stopped visiting us. Once I went to get some water, and"—her voice sounded almost incredulous—"Grandfather Shipunov knocked the pail right out of my hands, when it was full of water." Only now, more than half a century later, is Nikiforova again on speaking terms with the son of the family.

For two months after the police took away her father, as the weather turned colder, her mother took parcels of food and clothing to the prison. Later the family found out from a local NKVD man that all this had been in vain: Nikiforova's father had been shot some three weeks after his arrest. Each year now, on the day of his execution and on his birthday, she leaves flowers at the site of the grave, on the riverbank.

A stuffed owl stood in one corner of Nikiforova's living room. Potted plants lined the walls. Nikiforova's five- or six-year-old granddaughter, with black hair and a short summer dress, stood in a corner and listened silently as her grandmother talked. What was she taking in from all this? Was it anger at injustice, and hope that people will never again vanish in the night? Or was it a sense that the world is a place where men die young and end up as sepia photographs on a wall?

When the river revealed the bodies in 1979, Nikiforova said, "I went there the next day. The motorboats were forcing their wash against the shore. I saw a man whose body had not decayed. He was dressed in black and had a black beard. People at the Life-Saving Station told me that they had seen a woman with gold teeth, in a fur coat. I even think I know who that could be, the wife of the director of the Kolpashevo teacher training college. Both of them were arrested the same night." But no one saw Nikiforova's father's body.

"My son saw corpses floating along the river—and then they were sunk. It would have been so much easier to collect them. We have forgiven so many things. We could have forgiven this time again, and said, 'Well, let's take them all to the cemetery and bury them in a common grave.' Even the Nazis buried our soldiers."

How did the Kolpashevo massacre look from the other side, that of the secret police? In contrast to the large and faceless NKVD bureaucracies of big cities like Moscow, in a small town like Kolpashevo people at least knew who the other side was. One person they all knew was Stepan Marton. During the dread year of 1937, the year when Galina Nikiforova's father, the painter Gennady Sapozhnikov's father, and so many other people vanished into the hands of the NKVD, Marton was chief of NKVD regional headquarters in Kolpashevo.

Stepan Marton was originally a Hungarian—and, remarkably, a physician. Born in Budapest, he went to university there, and then served in the Austro-Hungarian forces in World War I. He was a prisoner of war in Russia when the Revolution broke out. Released from POW camp, he joined the Red Army, working as a doctor for a guerrilla detachment in the Russian Civil War. After the war, Marton remained in Russia. He joined the secret police, rose in its ranks, and married a woman who also worked for the police. Once he took up his post at Kolpashevo, part of his job would have been to sign the death warrants for those buried in the mass grave beneath NKVD headquarters.

On a pleasant Kolpashevo back street, green with the summer foliage, a grandmotherly woman nervously opened a door and invited us in. Beneath her tense smile, her face showed unusual intelligence.

Her name was Inna Sukhanova. She was Stepan Marton's daughter.

Like Galina Nikiforova, who lives a few blocks away, she too had worked all her life as a Russian literature teacher. "I love my work," she said, with feeling. The two women went to the same school, the one where Nikiforova's father was principal—although, uneasily, Sukhanova claimed she didn't remember him.

As Sukhanova told her story, it was clear that this cultured, thoughtful woman was still in agony about the events of a half century ago. For her, the pain was different than it was for Nikiforova, still mourning the death of a father she had loved. For the father Sukhanova had loved was an accomplice in the death, not only of Nikiforova's father, but of hundreds, perhaps thousands, of others.

It would have been far easier for Sukhanova had she not loved her father. But she talked about him with great warmth and admiration. A well-educated man, he spoke English, French, and German. He and his wife had three children; their son, Sukhanova's brother, was killed in World War II, and they adopted a war orphan. Her father, Sukhanova said, "was a gentle person, a man of integrity. He was not a malicious person. I always believed that he was an ideal, in all ways. He wanted very much for me to become a doctor." Intent on pleasing him, she followed his wishes and went to medical school. Only when illness forced her to drop out did she turn to her real love, literature.

Today Sukhanova still idolizes her father, but certainly not the regime he worked for. Lovingly framed in her living room were photographs of two of the great Russian counterculture figures of recent years, heavily censored before *glasnost,* the poet-singers Bulat Okudzhava and Vladimir Vysotsky. Sukhanova was painfully aware that in her youth, the Soviet Union had descended into madness. "Without any doubt there were lots of good people in the labor camps—honest, decent people who loved Russia. It was clear that people were arrested without reason. I know that now, and I knew it then. Arrested for *nothing.* Someone would wallpaper over

Inna Sukhanova holds a photo of her late father, NKVD commander Stepan Marton.

a piece of newspaper with Stalin's photo on it while he was remodeling his apartment, and suddenly he'd become 'an enemy of the people.'"

For much of her adult life, Sukhanova had clearly been struggling to reconcile this acute awareness with her love for her father, and she would continue that struggle, I felt, until she died. For her, that pain has not lessened with the passing of time, but has become more intense. For with each month that goes by now, more information comes to light or is released from police files. Just a few months earlier, the Kolpashevo newspaper, *Soviet North,* began running a column of thumbnail sketches and photographs of people executed at the old NKVD prison. This column appears in every issue. Fourteen hundred names have been released by the authorities so far; there are said to be many more.

Did Sukhanova's father have any doubts?

"I can't say," she replied. "I can only say that in our family, we never had a Stalin cult. We never owned a single portrait of Stalin.

All of us kids in the family were good students, and if one of us brought home such a portrait [handed out at school as a prize], Father would spend an evening making a certificate of good work and conduct to go under the glass, would make a border, and would slip it into the frame instead of the portrait."

Did her father—or mother—ever discuss his work?

"Never. They didn't talk about these things with us kids. We knew that we weren't allowed into Father's study, although it was never locked. We weren't supposed to go in, we weren't supposed to listen to the things being discussed there. That's how we were brought up."

Sukhanova knows how devastating the Great Purge was, because her father himself finally fell victim to it, one of some twenty thousand NKVD officers to be arrested. Marton was released after several years in prison. "I didn't sign anything," he told his family cryptically after he was set free, not saying much more.

Eventually, the NKVD took him back into its fold, although not quite all the way. For the first time in two decades Marton went back to work as a doctor; as a civilian, he ran *gulag* hospitals. "We were traveling from one camp to another," Sukhanova said, in Siberia and the Urals, living in the settlements for "free workers" near the camps themselves.

After Stalin's death, Marton was officially rehabilitated and restored to his old police rank. He then retired, and died in 1959.

Sukhanova had become more at ease as she was describing her father's imprisonment and all the pain that this put the family through; this was an ordeal shared by so many millions of others. It was obviously a relief to her to be able to talk about him in his role as victim rather than as executioner.

But finally we returned to the topic that had been hovering, unspoken, over our conversation: the mass of dead bodies in the bank of the Ob River, beneath the old NKVD prison. What did her father have to do with these killings? How could he not have known? Wouldn't he have had to sign the orders?

Sukhanova looked more anguished than ever. At first she tried to protect her father by including herself in the indictment. "You know, probably we all believed that there were 'enemies of the people.' . . . When those portraits of the dead appear in the newspaper today, I look at those lists, and every time I tell myself, 'Well, Father was in jail at that time, Father was not here at that time.' " Until quite recently, Sukhanova had found it easier to make such excuses to herself, but now her tone betrayed some doubt. And rightly so, for the death dates printed in the newspaper make clear that most of the executions at the Kolpashevo prison happened in 1937. Stepan Marton, Sukhanova's father, was not arrested until the very end of that year, when the family was already decorating the traditional New Year's tree.

She nervously began talking about something else, then interrupted herself and came back to the subject:

"His guilt. Probably his guilt exists, but it's a shared guilt. It's like blaming the soldiers, who fought just now in Afghanistan, for starting the war. The people who should be judged are those who *wanted* the war. The same applies here. When I came back here [to teach] in Kolpashevo, he said to me, 'Why are you going there? Mosquitoes. Cold. It may snow in July. Your lungs are weak. Swamps. Why are you going there?' I said, 'Papa, *you* went there.' He said, 'I was in uniform, I was ordered to go there. I couldn't refuse.' "

Listening, I wondered, why *had* she returned to Kolpashevo? It is just the sort of freezing, distant, provincial outpost that for centuries many Russians have desperately tried to get away from; Chekhov's stories are full of such people. Sukhanova implied that she remembered the town fondly from the happy days of her childhood, when the family lived here with an Alsatian dog named Dessie and her puppy, Caesar, before her father's arrest, before the frightening years he was in prison, before the long round of living near various *gulag* camps. But was she also, unawares, trying to atone for her father by returning to the scene of his crime?

I once saw a moving TV show about men and women in Germany today who were the children of high-ranking Nazis. In one way or another, they were doing penance for their fathers: one had become a priest, one a cabaret performer who sang the works of those killed in the death camps, one a human rights lawyer, and one—like Sukhanova—a schoolteacher. Sukhanova seemed driven by some similar impulse, although in her case, since she had tried for so long to deny the degree of her father's guilt, I think it was largely unconscious.

"Sometimes I criticized my father," Sukhanova continued. 'You—an old Communist. You saw what was happening. Couldn't anything be done about it? Why did you all have this psychosis in common?' "

Did Sukhanova know more than she was saying? Probably. But it was brave of her, I thought, to talk to me at all, and even more so to talk to Wilhelm Fast, who was from an organization seeking truth and justice for victims like the members of his own family. His frowning, gray-bearded figure must have felt to Sukhanova like an ambassador from the ghosts of all the people her father killed, the bodies in the riverbank.

Fast had been silent through almost the whole conversation; now it was his turn to question Sukhanova. "I could have asked you dozens of questions," he said to her intently, his gaunt face puckered with stress. "I should admit I have a heavy feeling. I sense that you don't feel comfortable talking to me. We belonged to such different worlds. But I hope that in the future we'll keep on talking. It's hard to ask questions. I feel that you love your father. But I would be afraid of him, because he was a *Chekist*."

"That's all right," Sukhanova said, trying earnestly to put him at ease.

"Your father had to carry out orders that came down from above," Fast went on. "If he hadn't, if he'd begun to resist, he wouldn't have worked there a month, not to mention a year." He paused. "It's scary for me to ask questions."

"Go ahead, ask them," said Sukhanova. She searched eagerly for some common ground. "Do you know why I didn't join the Party? I didn't see Communists as being the way my father was. That's why I refused to join the Party."

But Fast did not want to go off on this tangent. "How do you manage to live here?" he burst out, speaking a question that was on my mind, too. "Among all these people. How do people treat you? Here, now, today? Those who suffered directly."

"Few people know," Sukhanova said. "My name is Sukhanova, not Marton."

One who does know is Galina Nikiforova. She and Sukhanova have known each other since childhood. And even though the father of one almost certainly signed the death warrant of the father of the other, the two women are on friendly terms. Each had been in the hospital recently, and the other had visited. In their relationship, at least, they have put the past behind them.

Fast spoke about something else. Then he could not help himself from returning abruptly to the question of Sukhanova's father's guilt. "Excuse me, Inna Stepanovna ... To be in such a position, without being aware of things ... He *had* to have participated. I cannot imagine otherwise."

"I'm not trying to deny it," Sukhanova said sadly, speaking in double negatives. "I'm not trying to deny it. I don't think he had nothing to do with it." The way the Great Purge made everyone an accomplice, she said, was like a snowball, rolling downhill, picking up more and more snow in its inevitable course.

16
Shark
and Angel

What does one make of someone like Stepan Marton, man of culture and education turned mass murderer, doctor turned secret policeman, adoptive father whose killings created an untold number of orphans? His life seems to contain, woven into it, all the threads I had been following through the history of Russia in this century.

Marton had chosen to live in Russia. He was not the only former Austro-Hungarian prisoner of war, stranded in Russia, who joined the Red Army and fought in the Russian Civil War. Many hundreds of others did the same. But when the war was over, almost all of them went home. Marton stayed. Why? Whatever the attraction, it was strong enough to make him give up the likelihood of a comfortable life as a doctor in Budapest for another career, another language, and another, far poorer, country. And even for another name: he Russianized his Hungarian first name, Istvan, to Stepan. Yet he lived to see his dream of Utopia, if that is what it was, turn into endless bloodshed, with which his own hands were deeply stained.

To morally navigate through his life, and to avoid seeing what it had become, Marton had used with his daughter the most classic of denials: I was just following orders. But surely that rationalization alone was not enough. To be able to come home each day while a pile of corpses accumulated in the ground beneath his office must have taken every type of excuse there is. Furthermore, Marton knew some of these people. Could he really believe that the principal of his daughter's school, a fellow Red Army Civil War veteran, was a dangerous spy?

How much of a fissure was there in Marton himself? We don't know that his motives for staying in Russia after the Revolution were only lofty ones. Probably he was also attracted by the great life-and-death power of being a secret police officer. We are back again at the issue raised by the finger on the map. I want a railroad, *there*. Because it's good for humanity. Or, perhaps, because *I* want it there.

The most striking piece of evidence about Marton's divided soul is that this one-time signer of death warrants both started and ended his professional life by working as a physician. Of course not everybody who goes into medicine does so for noble reasons; after all, a doctor, like an NKVD man, gets to wield life-and-death power. Still, we are far more shocked than we would be if Marton's job before joining the secret police had been engineer or building inspector. How could a man be both healer and killer?

It is this duality that is at the core of the Russian Revolution, of the men and women who made it, and, at some level, of us all. In *Moby-Dick*, it is of this duality that Fleece, the black cook on board the *Pequod,* speaks when he leans over the ship's side and talks to the sharks who are tearing apart the carcass of a whale:

"You is sharks, sartin; but if you gobern de shark in you, why den you be angel; for all angel is not'ing mor dan de shark well goberned."

In his book *The Nazi Doctors*, Robert Jay Lifton looks at doctors who worked in the concentration camps; one, at Treblinka, actually served as a camp commander. Like Marton, most of these men were married, and most apparently had close bonds with their families. Also like Marton, many of the Nazi doctors who survived the war went back to practicing medicine. Dr. Eduard Wirths, who set up and supervised the "selections" system for the Auschwitz gas chambers, seemed to be remembered with particular warmth by his family. He, too, had a loving daughter. To a filmmaker who came to interview her, she said, "Can a good man do bad things?"

The answer, according to all too many studies of those who commit atrocities, whether they be Nazis, Americans in Vietnam, or South African police, is yes. If the enemy—Jews, blacks, Vietnamese, or "enemies of the people"—is sufficiently dehumanized by the relentless din of propaganda, if the murder or torture is sanctioned by peer pressure, by heads of state, by commands from those in authority, it is a rare person who will resist. Perhaps some people are born saints and some devils, but most of us are somewhere in between, influenced, in the end, by what the people around us are doing. It is the structure and customs and distribution of power of the society we live in that keep the "shark well goberned." Otherwise angels can, all too easily, be turned to sharks.

In trying to understand what allowed the Nazi physicians to let pressure from peers and superiors overcome professional ethics, Lifton speaks of the process of "doubling." It is an unconscious Faustian bargain, he says, a "division of the self into two functioning wholes." The concentration camp doctor had an "Auschwitz self" which he needed to act as part of the death machine, and a "prior self," which took no responsibility for anything done at Auschwitz. "The feeling was something like: 'Anything I do on planet Auschwitz doesn't count on planet Earth.'" One does not have to be a concentration camp officer to practice this doubling, Lifton points out. An army psychiatrist doubles when he certifies a soldier as fit to be sent back to a job where he can kill and be killed; a nuclear physicist who loves his wife and children doubles when he works on an atomic bomb. And, we might add, the issue does not only arise with war and weapons; a kind-hearted, churchgoing engineer doubles when he allows a new product off the assembly line without adequate safety testing. The list could go on.

After spending the day in Kolpashevo, Wilhelm Fast and I flew back to Tomsk, the city of the old wooden buildings. Fast's bag now held some cucumbers from his mother-in-law's garden. His long frame crammed into a tiny Russian taxi for our trip into Tomsk from the airport, Fast was still troubled by our meeting with Inna Sukhanova. You can't blame the children for the sins of the fathers, he said, but "my relatives were shot. Her father ordered people shot." A deeply religious man, he was clearly torn between a belief in forgiveness and his horror at meeting a mass murderer's daughter. The two feelings were at war in his face.

There was one more person I wanted to see before I left Tomsk, someone who might have something to say about that mass of bodies in the riverbank, and perhaps about Stepan Marton's role in the killings. In the last few years, a half dozen or so dissident former secret policemen have emerged from the shadows and gone public. And one, I had heard, was right here in Tomsk—a veteran of service in the region that includes Kolpashevo.

Anatoly Spragovsky is a short, overweight, red-faced sixty-six-year-old. He joined the secret police in 1947, he told me. His first assignment was to Moscow, to shadow foreign diplomats as they went about the city—"sitting on their tails" is the Russian expression. Then he came back to his native Siberia, to the *oblast* that includes Tomsk and Kolpashevo, and worked here until 1960. Although he entered the police a decade after the Kolpashevo massacre, he was, he said, "a sort of indirect witness to the atrocities of that time, because I've studied the documents."

So far, no one outside the KGB has seen these documents. Spragovsky had seen them because for most of the 1950s he was—like Nikolai Danilov, whom I had met in Moscow—a rehabilitations investigator. In that period, each time a living or dead person was to be rehabilitated, the authorities formally reopened the case. An investigator like Spragovsky examined papers, visited the place where someone had been arrested, reinterviewed surviving witnesses, and took statements from police officials. Then a board solemnly declared that the man or woman involved was not guilty, after all, of being a counterrevolutionary, an industrial saboteur, a Japanese spy, or whatever.

Rehabilitations in the 1950s proceeded very slowly, one case at a time. Someone once asked Anastas Mikoyan, an adroit survivor who served in the Politburo both under Stalin and long after, why this was so. Couldn't all these myriad "enemies of the people" simply be declared innocent all at once? "No they can't," he replied. "If they were, it would be clear that the country was not being run by a legal government, but by a group of gangsters." He paused, then added, "Which, in point of fact, we were."

The fiction that the millions of Great Purge arrests were a string of individual mistakes required each case to be laboriously proven a miscarriage of justice. Because of this, a vast amount of 1930s top-secret paperwork passed through the hands of rehabilitations investigators like Spragovsky. Spragovsky also questioned guards, executioners, and other secret police officers who ran this machinery of mass murder, to get them to testify, for each case, that they had fabricated the evidence. From all this, he could piece together an extraordinarily vivid picture of the Great Purge from the inside. He knows the kind of details rarely, if ever, glimpsed by prison survivors and historians.

Before people were even picked up in the mass arrests of 1936–39, Spragovsky explained, "everything was planned in Moscow. On the walls [of NKVD offices] were charts of various counterrevolutionary organizations." These organizations were, of course, totally imaginary. But arrested people had to be charged with being agents of something. In Moscow or Leningrad, people were often accused of being spies for some foreign power; engineers were charged with sabotaging the factories where they worked; and so on.

But for rural, western Siberia, with few factories or nearby foreign powers, this wouldn't do. Instead, it was decided that the major enemy was to be a group "called the Counterrevolutionary Cadet Monarchist Organization. They decided to declare all prisoners who had been arrested members of this organization.

"This group was the fantasy of the central NKVD. The underlying principle of this Counterrevolutionary Cadet Monarchist Organization, according to their design, was that it was like an army. The entire region was said to be infested with this army. People were charged with belonging to this or that military unit. If it were people from Kolpashevo, which you visited, then they were charged with belonging to the Kolpashevo regiment."

This was consistent with what I had heard from the painter Gennady Sapozhnikov, unexpected witness to the uncovered Kolpashevo grave, whose executed father was indeed charged with being a monarchist.

Anatoly Spragovsky's potbelly, bald head, and short-sleeved summer shirt gave him the look of a retiree in Florida. While a warm June breeze blew through the window of a Tomsk apartment, he leaned back on a couch and talked in a loud torrent, often temporarily ignoring a question in order to keep on answering the preceding one more exhaustively. His account bristled with names, dates, and articles of the legal code. At first, one might think his was just the voice of a stubborn, elderly person beginning to get a bit deaf. But it was also, I began to see, the voice of a brave man who had made it his business to remember everything and forget nothing. And who had been wanting to share it all for a long, long time. He would have gladly talked all afternoon, it seemed, and into the night.

"The interrogators were violent," Spragovsky went on. "The main aim was to make a prisoner sign his interrogation record. Most of these people were illiterate. And so they stamped the prisoner's finger on each sheet. Those who refused had their fingers squeezed between the door and the wall until their fingers were broken.

"What were the specific crimes they were accused of? In Kolpashevo, for example, of a plan to blow up a bridge over the Ob River. But at the time this bridge did not exist! In Tomsk, they would charge people with planning the explosion of a bridge over the Tom river, which also did not exist."

The Purge did not have to do, after all, with punishing people for actual crimes. To the extent that it had a conscious purpose, it was to inspire terror and obedience. The very impossibility of the crimes must have added a spooky dimension to that terror. To be charged with plotting to blow up a bridge that did not exist must

Secret police whistle-blower Anatoly Spragovsky.

have made people feel at the mercy not only of a cruel and powerful state, but of a frightening, incomprehensible new system of logic, against which there could be no appeal, no argument, no demonstrating that the crime you were accused of was impossible. This must have been what the helpless women accused of witchcraft felt, as they were charged with casting spells or turning humans into bats.

Why did the NKVD try to implicate everyone in giant conspiracies? Stalin's biographer Alex De Jonge suggests that these imaginary plots were a mirror, in Stalin's mind, of a real one. For was he not, at this very moment in history, planning his secret agreement with Hitler to take over all the countries that lay between them? Even in these remote Siberian villages, we may be seeing, projected onto others, Stalin's own fear of persecution for what was—in degree of secrecy, and in the number of people affected—one of the greatest conspiracies of all time.

Like the rest of the Soviet economy, the entire Purge operated by a system of quotas. But Spragovsky stressed as he continued his description that these quotas were not limits, they were minimums. He explained: "In the documents there were orders with exact numbers given. You would be ordered to arrest and shoot, say, ten thousand people. But you could arrest twenty, thirty, forty thousand. The more 'enemies of the people' you arrested, the higher your score. As a result of such orders, the Tomsk NKVD challenged Novokuznetsk to a socialist competition. The cities were the same size. Who could arrest more people? It turned out that Tomsk arrested and executed more people than Novokuznetsk."

The Tomsk authorities did indeed win the contest hands down; by some estimates, the NKVD shot roughly one out of every six residents of the city. The Tomsk NKVD boss won an award for this showing, Spragovsky said, and then added: "The same was true for Marton. For the Kolpashevo operation, they gave him a medal. I believe it was the Order of the Red Banner.

"Marton's guilt goes without saying. I've seen all the documents and papers connected with him. He was one of the people who organized all these illegal repressions. There's no doubt about it. Some three or four dozen employees who were falsifying all these cases reported to him. I interviewed Marton's deputy."

Spragovsky does not, however, blame Stepan Marton any more than he does anyone else. "If he hadn't done all this, he would have been shot." When he was working on rehabilitations in the fifties, Spragovsky said, he went over cases with other officers who, like Marton, had sent many people to their deaths. "When I presented them with proof that they had fabricated a case, they wouldn't even bother to look, since they already knew that everything was a lie. But they'd say to me"—and here Spragovsky used the more direct and intimate second-person singular—" 'If you had been in our shoes then, you would have done the same.' "

You would have done the same. In some ways, this is the most unsettling thing about the Kolpashevo killings, and about the entire paroxysm of violence that swept the Soviet Union in those years. We want so hard to believe that heroic resistance to evil is always possible, that there still can be angels in a world run by sharks. After all, what about Gandhi, Martin Luther King, Jr., Nelson Mandela? Of course you can always resist evil. But to resist and have a chance of surviving is a different matter. To do that you either have to have a government that doesn't immediately murder anyone who protests, or there have to be enough of you so that the regime is afraid to kill you all. Neither of these conditions existed in the Russia of the late thirties. By the time the Great Purge began, it was too late for dissent, too late for resigning in protest, unless you were ready for instant death. As it happens, one of the tiny handful of NKVD commanders known to have tried to slow the murder machinery was stationed near Tomsk. His name was Dolmatov; he refused to make some arrests and he reopened a closed church. He was promptly shot. The NKVD had no room for conscientious objectors.

You would have done the same. Anatoly Spragovsky was unflinchingly honest. He himself, caught in the same deadly system, *had* done the same, or close to it. How many times I don't know, but he gave one example from his early days in the secret police, as a true believer, right after Stalin's death. "We were all convinced that we were surrounded by spies, and that we couldn't find them all until Doomsday. Our task was to uproot this scum. We believed this. I was involved in several cases like that. For example, the case of Ryabov. He had climbed onto a pile of logs and shouted, 'Down with Bulganin and Khrushchev! Long live Eisenhower!' He said this in the presence of some twenty people. He was just a backwoods guy—out in the *taiga,* cutting timber, eaten up by mosquitoes. His work gang was on smoking break. Somebody informed on him. I investigated the case. I interrogated everyone, and they all confirmed the story. Ryabov himself did not deny it. Everything went according to protocol. Our verdict was that he was an enemy. He got ten years."

Someone else who had done the same, Spragovsky discovered in the police files, was his own father. "He was the chairman of a rural council. It turned out that in 1938 he had helped the NKVD. The NKVD composed its lists [of people to be arrested] through the rural councils. My father was among those who helped pick these people.

If he hadn't, he would have been jailed himself. That's how he saved his skin. He also had five brothers, and he managed to keep all his brothers alive by collaborating with the NKVD."

What exactly happened at Kolpashevo? What did Stepan Marton get his medal for? Spragovsky's answer was again consistent with the story of the painter Sapozhnikov, whose vanished father had last been seen heading toward Kolpashevo on a prison barge.

"The Ob River runs through this region, and the Tom River is its tributary. The authorities took two barges. From the upper reaches of the rivers, they started to collect people. When they had picked up all the people arrested at each place, they moved farther along. So they collected several thousand people in the two barges, and brought them to Kolpashevo, and anchored the barges in the middle of the Ob. Why? Efficiency—the prison wasn't big enough to hold everybody.

"There were no public trials. Nobody could appeal. Those arrested were executed at the very spot, within the walls of the [Kolpashevo NKVD] headquarters. Sergei Karavaev, who was on the team that did the job, told me how. They had a special group of executioners. Inside the walls, they dug a pit, a deep one. There was a wooden walkway leading to the pit. When a man approached this pit, he didn't know he was going to be shot. He was walking forward, unaware that there was a pit in front of him, because there was a sort of curtain hanging there.

"The [executioners] stood to one side. When a person came to the edge of the pit, the bullet was fired. As a rule they shot people in the back of their heads. Then the person would drop into the pit." This pit, then, was the mass grave that the Ob had torn open some forty years later.

In other cities, Spragovsky said, executioners sometimes used a method called "soap and rope": "The pretext was to save bullets. The executioners would soap a rope, throw it around the victim, and hold it tight for some time. After a while the person strangles. Another episode: in Kuybyshev, in the Novosibirsk region, two schoolteachers were in love. They were young. When the sentence on them was to be carried out, the executioners were drunk. They said, 'If you copulate with her with your hands tied, we'll save your lives.' They put him down on the other teacher, inserted his organ into her. He started. Then they threw a loop of rope around them both and strangled them."

After some years of rehabilitating Great Purge victims, Spragovsky grew angry that so many people who had carried out these arrests and executions were still, in the late 1950s, in high office. He began arguing with his superiors about this. As he described these clashes to me in dense, bureaucratic detail, studded with references to NKVD rules and departmental assignments from long ago, it struck me that Spragovsky's personality was like that of several U.S. government whistle-blowers I've interviewed over the years. You expect that men or women courageous enough to speak out against injustice at great risk to themselves are doing so from a belief that human ethics transcends all laws. Instead, they are often driven by something slightly different: a passionate, legalistic outrage that the Pentagon, the CIA, or whatever, is doing something *against its own regulations.*

In 1959, frustrated, Spragovsky wrote a letter to Khrushchev. He urged him to stop the nonsense of case-by-case rehabilitations, and to declare all victims of the Great Purge innocent in one swoop, by national law. Then he stood up in a Party meeting at the KGB, he said, and demanded that "former falsifiers" who were now in high positions be ousted. He was fired. For a long time the KGB prevented him from finding other work by warning prospective employers not to hire him. Eventually he found a low-profile job in a gear-repairing workshop. "For twenty-seven years I did not, so to speak, raise my head." Now that he can do so at last, he has raised his head again.

As I think back over the deep wound in so many people's lives that the spring floodwaters of the Ob reopened on that day in 1979, the person whose story haunts me most is Stepan Marton, the Budapest doctor who ended up as a secret police commander in Siberia.

Although he won a medal for committing his quota of mass murder, was Marton an evil man to begin with? It would be easier if he was. Otherwise we are faced with the chilling possibility that history can force any of us into situations where *you would have done the same.* But that is, I fear, the case. Come back for a moment to the world in which the young Stepan Marton finds himself, as a newly released prisoner of war, in 1917 and 1918. All Europe is horrified by the unspeakable, needless carnage of World War I—which Marton, of course, has seen first hand. The old empires, Russia, Germany, his own Austria-Hungary, are crumbling. In deciding to support the Russian Revolution, Marton is not alone: thousands of other foreigners also fight on the Red side in the Civil War, the first battle in building the society that will end war and injustice forever. Time is short; the forces that will soon thrust Hitler and Mussolini to prominence are already on the horizon. It is a world in which it is easy to feel, *There must be a better way.*

Many of these foreigners in the infant Soviet state join the Cheka, the first incarnation of the secret police. Its first chief, Felix Dzerzhinsky, is a Polish nobleman who had spent many years in Tsarist prisons; a man who eventually becomes a top general is a Turkish Jew; the future chief of Stalin's bodyguard is, like Marton, a former POW from Budapest. The Cheka's ranks include many other Jews and Hungarians, plus Finns, Latvians, and Czechs. It is an elite force at the very forefront of the battle for the new society. Yes, yes, the Cheka's role is to quash dissent, but look, the threats to the Revolution are so huge, we're surrounded on all sides. Foreign troops—British, French, American, Japanese—are on our very soil, shoring up the remnants of the White armies that are trying to overthrow the new state, the first in all history controlled by workers and peasants. Churchill has called on the West to "strangle the babe of Bolshevism in its cradle." Spies—real ones—are everywhere; the whole world is trying to crush us. Who can now afford petty peacetime luxuries like free speech, jury trials, opposition parties, carping critics? Maybe later ...

Even Victor Serge, an uncompromising opponent of the secret police and all it stood for, acknowledges that at the beginning the Cheka was headed by "incorruptible men like the former convict Dzerzhinsky, a sincere idealist, ruthless but chivalrous.... But ... gradually ... the only temperaments that devoted themselves willingly and tenaciously to this task of 'internal defense' were those characterized by suspicion, embitterment, harshness and sadism." We do not know where along this spectrum Stepan Marton fell, nor perhaps how he changed from the one type to the other.

In talking about post-revolutionary history with Russians, I often asked when they thought the Soviet Union had turned its back irrevocably on the paths not taken. After what point did the great famine, the Purge, the *gulag,* and all the rest of it become inevitable? After the Revolution itself? After the Bolsheviks suppressed the other political parties? After Stalin became General Secretary of the Communist Party? One can ask the same question about the life of someone like Stepan Marton. What was his point of no return? As a prisoner of war in a senseless, bloody conflict between two dying empires, Marton must have felt himself to be an innocent victim. When did he turn guilty? Just as a small act can be the essential step on the course toward great heroism—Susanna Pechuro and her friends refusing to read their poetry aloud at the officially sponsored youth literary club—so complicity often starts with something small. A border is crossed, and then it is hard to turn back. Did Marton make this step when he took part in the Russian Civil War? Joined the secret police? Took up his post in Kolpashevo? At what point was he trapped?

There is another question about Marton that I don't know the answer to. After his own release from prison, in what spirit did he go back to work as a doctor in the *gulag?* Was there some element of atonement—did he want to spend his last working years taking care of prisoners as a physician instead of signing their death sentences? Or, more likely, was it because this was the closest thing he could find to his old job with the NKVD? It would be nice to believe, of course, that his motive was more the first than the second. We always want even the worst villains to feel at least a hint of repentance. But I do not know if he did, and I suspect that even his daughter, loving him still, despite her anguish, does not know. It is one of the many secrets that the man who presided over the Kolpashevo killings took with him to his own grave.

THE DARK SIDE
OF THE MOON

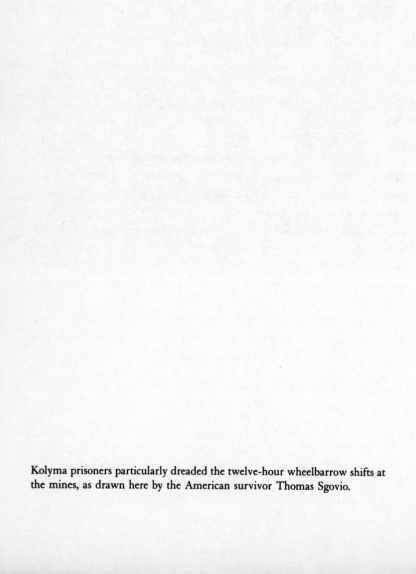

Kolyma prisoners particularly dreaded the twelve-hour wheelbarrow shifts at the mines, as drawn here by the American survivor Thomas Sgovio.

17

Beyond
the Pole Star

One last place remained to see, the final destination of my trip across Russia to the northeast corner of the Eurasian landmass: Kolyma.

Centuries before Kolyma became the site of the most deadly camps of the *gulag,* legends about the territory's coldness and remoteness reached the outside world. They were even heard by Marco Polo, as he was crossing the continent far to the south. Judging from some directions he gives, Polo is referring to Kolyma when he writes: "You must understand that there are neither men nor women here, nor beasts nor birds. . . . I assure you that this region is so far north that the Pole Star is left behind towards the south."

Soviet prisoners sent to Kolyma nearly seven hundred years later felt that what they were leaving behind to the south was the change of seasons. One of their songs goes:

> Kolyma, Kolyma,
> Wonderful planet!
> Twelve months winter,
> The rest summer.

Most of Kolyma is encompassed by the watershed of the Kolyma River, which snakes and tumbles through this rugged region, the last of the great north-flowing rivers reached by travelers heading from Europe across Siberia. The early explorers found a thin scattering of native peoples in the area, subsisting on fish, bear, and reindeer like their Eskimo cousins across the Bering Strait. But the climate was too inhospitable for the indigenous population to grow large. With some mountains reaching 10,000 feet, parts of Kolyma are colder than the North Pole.

The short story writer Varlam Shalamov describes how the white frost of Kolyma covered everything, even death warrants:

> For many months there day and night, at the morning and the evening checks, innumerable execution orders were read out. In a temperature of fifty below zero the musicians from among the non-political offenders played a flourish before and after each order was read. The smoking gasoline torches ripped apart the darkness.... The thin sheet on which the order was written was covered with hoarfrost, and some chief or other who was reading the order would brush the snowflakes from it with his sleeve so as to decipher and shout out the name of the next man on the list of those shot.

Bounded by the Arctic Ocean to the north and a gulf of the Pacific—the Sea of Okhotsk—to the south, and by ranges of rocky, snow-covered peaks on each side, the Kolyma basin is a vast natural prison. It was so used as early as the seventeenth and eighteenth centuries, when members of the Tsar's guard found guilty of conspiracy were exiled there. Even today Kolyma is difficult to reach by land. The one road into the territory is closed by weather for much of the year; in the few months that trucks can get through, the drive from Moscow takes twenty-four days. This road didn't exist for the first decade of the Kolyma *gulag*. Permafrost and the mountains have defeated several tries at building a railway. It was only after World War II that Kolyma was even linked to the south by a telephone line; until then, communication with "the mainland" was by short-wave radio.

For Kolyma's prisoners, escape was out of the question. The only known successful attempt was by a man who made a months-long trek over the mountains, mostly on foot. He posed as a geologist, checking in with the authorities in towns and collective farms along the way, sometimes asking people to help him by mailing packages of rocks back to the Academy of Sciences in Moscow. He was finally caught, but only a year or two after he had arrived back on "the mainland."

Even in the other parts of the *gulag,* Kolyma had a special reputation, because so few of the prisoners sent there returned. Michael Solomon, a Romanian, was in a labor camp on "the mainland" when he was assigned to a convoy headed for Kolyma. "When the time came to leave," he writes, "all the camp inmates accompanied us to the main gate, exactly as mourners at a funeral."

Secret police authorities in Kolyma today say there are records—sometimes a complete file, sometimes just a name on a list—of two million men and women who were shipped to the territory between 1930 and the mid-1950s. But no one knows, even approximately, how many of these prisoners died. Even historians who have spent years studying Kolyma come up with radically different numbers. I asked four such researchers, who between them have written or edited more than half a dozen books on the *gulag,* what was the total Kolyma death toll. One estimated it at 250,000, one at 300,000, one at 800,000, and one at "more than 1,000,000." The documents that could tell us more are almost certainly destroyed. We will probably never know the answer.

Many prisoners died before they even got to Kolyma, for each stage of the long journey there was an ordeal in itself, claiming tens of thousands of victims. And like stations of the Cross, each stage was branded into the memories of survivors. Can it become worse than this? Yes, it can. The first stage was the long trip across Russia on the Trans-Siberian Railroad, where the inmates were packed into crowded boxcars. One prisoner, Vladimir Petrov, was locked in a railway car for forty-seven days, from Leningrad to Vladivostok. Like most prisoners on this journey, he was given very little water. Prisoners felt its lack keenly because the meager amounts of food they got were often salted fish. Inmates wrote letters and messages on scraps of paper and tossed them out of the train, hoping someone would find and mail them. In winter, a layer of frost several inches thick formed on the inside of the boxcar walls. The bodies of those who died traveled on, frozen, among the living.

Next, at the Pacific end of the railway line, the prisoners were locked up, sometimes for months, in huge "transit camps." These were dominated by thieves, who robbed newly arrived convicts of anything of value. From the transit camp at the port of Vladivostok, awaiting shipment north to Kolyma, the poet Osip Mandelstam sent a last, desperate note home: "I'm in very poor health, utterly exhausted, emaciated, and almost beyond recognition. I don't know if there's any sense in sending clothes, food and money, but try just the same. I'm freezing without proper clothes." It was December, and the sea route to Kolyma was frozen over for the winter. Mandelstam died in the transit camp. His body, like those of other prisoners, lay unburied until spring, when the ground thawed enough for them to be thrown into a common grave.

For those who survived this far, there came the week-long sea journey to Kolyma, often in a vessel from the large fleet, with a band of blue paint on the funnels, controlled by the NKVD itself. In 1933, the *Dzhurma* sailed from Vladivostok for Kolyma too late in the season, and got caught in the ice for the winter. The ship, its crew, and the NKVD guards on board survived; the several thousand prisoners below decks did not.

Prisoners' descriptions of the sea trip to Kolyma evoke that other great slave voyage, the Middle Passage across the Atlantic: the crowded holds that reeked of disease and feces, the great storms at sea. But unlike the African captives, Kolyma-bound prisoners were not chained. There was no need; they could be easily disciplined with streams of seawater from fire hoses, a deadly threat in below-zero weather. In the ships, too, the criminals ruled over the "politicals." "We lay squeezed together on the tarred floor of the hold because the criminals had taken possession of the plank platform," writes Elinor Lipper, a Dutch-born Kolyma survivor. "If one of us dared to raise her head, she was greeted by a rain of fish heads and entrails from above. When any of the seasick criminals threw up, the vomit came down on us."

Thomas Sgovio, the young American who had been in the same Moscow prison as Arthur Talent, was transported to Kolyma on the ship *Indigirka*. When it passed near the coast of Japan, the hatches were tightly covered, so that no Japanese fishermen would suspect any human cargo was aboard. In the resulting complete darkness, where prisoners could talk without fear of being identified by informers, Sgovio heard anti-Soviet remarks for the first time.

Peter Demant, a former Kolyma prisoner I met in Moscow, described how inmates on the ship he traveled on were let out once a day to use a toilet. This crude structure consisted of two long planks with a space between them, and a roof, slung over the ship's side. "And there the prisoners sat." One day, "there was a big storm. And in one fine moment, a wave carried away the hut, the planks, and the people—about seven or eight of them."

After six to eight days at sea, the ships would dock at Nagayevo Bay, the harbor of Magadan, Kolyma's capital and main port. When Michael Solomon disembarked there, all five thousand prisoners on his ship were drawn up in ranks on shore. The scene he describes is almost phantasmagoric. NKVD General Panteleimon Derevenko, the ruler of Kolyma, rode up on a white horse. He spoke to the shipload of inmates in a loud and echoing voice:

"Convicts! This is Kolyma! The law is the *taiga* and the public prosecutor is the bear. . . . You are here to work, and to work hard! You must repay with your sweat and tears the crimes perpetrated against the Soviet state and the Soviet people! . . . The fulfillment of the Plan is our sacred duty. Those who do not fulfill the Plan are saboteurs and traitors, and we show them no mercy!"

Surviving prisoners were then marched off to be distributed through the more than one hundred camps of the vast Kolyma system. At the transit camp in Magadan itself, writes Solomon, "We were paraded naked in front of a medical board of young girl interns just out from the Moscow University Medical School. Then we were carefully examined by broad-shouldered unfriendly-looking civilians dressed in sheepskin coats and wearing high felt boots. These were the mine managers. . . . They pinched our muscles, opened our jaws, and felt our teeth. If satisfied, they addressed the board with the words, 'I take him.'"

Prisoners were expendable labor, nothing more. "Your lives are reckoned to last ten years," an NKVD doctor told Solomon. "If you live longer, it means that you are guilty of one of two things: either you worked less than was assigned to you or you ate more than your proper due."

My visit to Kolpashevo and Tomsk had taken me nearly half the distance from Moscow to Kolyma. I made one more stop in central Siberia. Then I boarded a plane that was supposed to fly overnight and land me in Magadan the next morning. But the Magadan airport was closed off by a Pacific fogbank, and in the middle of the night we came down to wait for five hours at Yakutsk. Other stranded planes from all over the Russian Far East had landed in Yakutsk, and the airport was jammed with people: Russian, Yakut, shaven-headed young naval recruits, civilians—awake, asleep, drunk, sober, walking around, or sprawled across coats and suitcases on the floor. The airport men's room was a dim cellar containing only, on all four sides, a row of holes in the floor.

The plane took off again at dawn, and by midday I was in Magadan, where at the airport a familiar face was waiting for me: the head of the local Memorial branch, Miron Markovich Atlis. He was sent to the *gulag* in the early 1950s because he had been friends with a member of Susanna Pechuro's anti-Stalinist group of teenagers. Atlis was the man who, some forty years later, called Memorial headquarters in Moscow, found himself talking to Pechuro on the phone, and exclaimed, "Susannochka, you're alive!"

I had met Atlis, now a sixty-one-year old psychiatrist, when he was on a visit to Moscow a few months earlier, and although I was later to feel very differently, I have to admit that my first impression had not been a good one. He was short and heavyset, with bulging eyes, thick spectacles, and a deep, stentorian voice. I had seen him at his son's apartment. "Papa, pipe down! Pipe down!"

his son kept saying. Furthermore, Atlis had very definite ideas about whom I should and should not talk to once I came to Magadan. He had said I should warn him when I was coming and he would arrange everything.

Instead, I had decided to arrange things from Moscow, primarily visits to some of the old labor camps themselves. This was not easy. One of the Soviet economy's chronic bottlenecks was building supplies, and so the hundreds of *gulag* camps shut down after Stalin's death have almost all been dismantled by people scavenging for lumber, bricks, glass, or anything else that can be carried off and used. The only camps still intact are those no one can reach by road. Such places I would have to visit by helicopter. One of these was available in Omsukchan, a town some 200 miles northeast of Magadan. A man there had recently set up a bear hunt for several adventuresome foreign diplomats who had flown out from Moscow, and he would make the helicopter arrangements.

And so I had waited until the day before I was to arrive in Magadan to phone Miron Markovich Atlis. I figured I would pay a short courtesy call on him in his role as head of the local Memorial group, and that would be that. But when I reached Atlis on the phone and started to explain that I was the American he had met several months earlier, he interrupted: "Of course! I know what plane you're on, and I'll be at the airport tomorrow to meet you! And our trip up to Omsukchan is all arranged."

Our trip? It turned out that the bear-hunting contractor, faced for the first time with someone hunting prison camps instead, had figured he needed help. He had turned, quite naturally, to the nearest chapter of Memorial.

There was no escaping Miron Markovich Atlis. He was there on the tarmac when I got off the plane at Magadan, puffing on a cigarette, his voice booming, his eyes bulging, his bulky body plowing a path for us through the airport crowds. On the hour-long bus trip over bone-cracking roads into town, he kept up a non-stop commentary on the flora and fauna of Kolyma, the town's industries, and the old *gulag* barracks we could see from the road, now used as warehouses. After the interrupted overnight flight, I was exhausted. At my hotel, he insisted on coming up to my room to make sure everything was all right.

"And now," he said, "it's time for lunch. And we'll plan our afternoon program." He grabbed me by the arm and led me into the hotel restaurant.

Miron Markovich was a great arm-grabber. At lunch he reached out and seized the arm of someone walking past our table, introducing him as an official of the city government. As private cars are even scarcer here than elsewhere in Russia, Miron Markovich asked him for a city car, so that this "very important foreign visitor" could see the sights of Magadan. Yes, the official said politely, the city would provide a car. Like me, he seemed to feel in the presence of an inexorable force.

Some time later, Miron Markovich grabbed the arm of a mournful-looking young woman with a black bow in her hair, a friend of his daughter's, who had come to his apartment to return a book. "Come!" he commanded. "You must talk to this American journalist." When he had seated us together, he ordered me, "Turn on your tape recorder."

He was right. The young woman was studying history at the local teacher training college. But she was deeply disillusioned with the rigidity of the educational system. As she gave examples, and made comparisons with pre-revolutionary Russia, Miron Markovich paced back and forth approvingly, blowing out quantities of smoke, prompting her when her answers grew hesitant, and prompting me when he thought I didn't ask the right questions.

That afternoon, Miron Markovich commanded, "Now, show me that list of people you're planning to see." Weakened and punchy from sitting up most of the night in the Yakutsk airport, I handed it over.

"Hmm," said Miron Markovich, running his finger down the list, blowing smoke, pausing at each name. "Okay ... Okay.... This fellow doesn't know anything I don't know; you can skip him ... Okay ... This guy'll be too drunk to talk to you.... Okay ... This one's unreliable; we can find you someone better than him.... This one—maybe.... This one's not worth talking to, nothing to say.... This one's out of town.... This one we can go see right now...."

"Wait!" I protested feebly. But Miron Markovich was already making arrangements. At last—still escorted—I managed to get back to the hotel. There he insisted on coming up to my room, to help me make phone calls to set up the remaining appointments. Not necessary, I insisted. "But it is!" he said. He was right. Phone numbers change unpredictably in Russia, and half the time I turned out to have the wrong ones. Miron Markovich seized the phone like a weapon, boomed, "Miron Markovich here!" when people knew him, and, "Atlis, Miron Markovich!" when they didn't, battled recalcitrant information operators, spouses, and secretaries, and managed to get most of the people I wanted to reach on the phone, handing over the receiver to me each time with a gleam of triumph in his eyes.

Making the rounds of my appointments the next day, I realized that, despite everything, I was learning a good deal from Miron Markovich. He possessed a vast fund of information about the *gulag* and Kolyma, and a passion for sharing it, at lecture-platform volume, with anyone in earshot. Putting together what I had read with what he showed me when we drove around town in the city car he had commandeered, I got a much better sense of this small, wind-swept city whose name alone to Russian ears has the sound of doom.

Some half a million people live in Kolyma today, of whom about 160,000 are in Magadan. There was not even a small settlement at Magadan until the 1920s, when prospectors realized what great mineral wealth lay beneath Kolyma's snow and ice: platinum, uranium, coal, and, above all, gold. The Soviet government began exploiting the region in earnest at the start of the 1930s. Convicts were shipped to Magadan, marched into the countryside, and made to build their own camps. Supplies lagged far behind. In some camps there were no mess tins, and prisoners had to drink their watery soup from their caps, from curved pieces of bark, or from their boots. By the 1940s, Kolyma was producing, by some estimates, nearly one third of the world's gold.

All miners were prisoners, underfed and ill-clothed, forced to tramp or climb miles to and from work in the frigid polar darkness. They suffered particularly after 1937, when Stalin made a speech accusing labor camp administrators of "coddling" convicts. Anxious to save their own skins, *gulag* authorities banned convicts from wearing fur clothing or felt boots. This meant facing the Arctic winters with wadded jackets and footwear of canvas or sacking. By one estimate, each ton of Kolyma gold cost roughly one thousand human lives.

Magadan was built by convicts. Prisoners paved its streets, poured its concrete, and, in the nearby forests, cut the wood used for construction and heat. Today the trees on the hills around Magadan are low and sparse, like the hair on a balding scalp. The original forest was all stripped to build the city, Miron Markovich explained. General Alexander Gorbatov, a purged Red Army officer, writes of cutting trees on the outskirts of Magadan and dragging them out of the forest on sledges:

> Picture a straggling line of men, or rather shadows of men, strung out over three miles as they weave their way along the mountain slope, gaunt, draining their last ounce of strength, their necks stretched forward like cranes in flight as they drag the wood behind them. It is hard enough to pull a load downhill, it is harder still to pull it on the flat, but the slightest slope puts it beyond your strength: you stumble, fall, get up, and fall again, but nothing will shift the load except another man pushing from the back. Somehow we got the wood back to the camp.

Although its time as a *gulag* capital is now passed, Magadan still has a rough-and-ready feel about it. Many of the people who live there have come only for five or ten years, to earn "the long ruble"—the extra pay you get for working in the Far North. Every other vehicle on the street seemed to be a truck or jeep or bulldozer. A few years ago, a bear wandered into the center of town and had to be frightened away by police gunshots. Living in Kolyma is not for the faint of heart; even today, more than one third of the region's housing units do not have indoor toilets—and outdoor winter temperatures here routinely fall below -60° F.

Even more so than in other Russian cities, piles of debris lay about the streets of Magadan—chunks of cement, broken bricks, tangled metal rods and pipes, rusted oil drums, half-assembled scaffolds. Many buildings had that distinctively Soviet look, where you cannot tell whether something is being built or dismantled. Thick, wind-blown fog from the ocean covered the city. Enormous crabs, claimed as the Pacific's largest, were on sale from stands along Magadan's sidewalks. Because crab sellers did not offer bags, women carried the crabs home holding onto one massive claw.

Although it was mid-June, patches of snow lay on the rocky hillsides, and men and women on the streets wore thick overcoats and fur hats. Because it was officially summer, however, central heating was turned off. Apartments were warmed by small plug-in electric heaters, but there were none of these in the hotel. After inspecting my clothing, Miron Markovich boomed at me with doctorly disapproval, "You won't be warm enough!" As usual, he was right. He loaned me a thick extra sweater for the rest of the week.

Several people in the city told me wistfully that they hoped to attract tourists from Alaska: "We're so close!" It will be a while, however, before the Hotel Magadan earns any stars from the Mobil Travel Guide. My room swarmed with cockroaches, and the hot water came out of the tap dark brown. When I dutifully tried to call the Foreign Ministry in Moscow to report my whereabouts, as required by the rules then in effect for foreign journalists, I was told that all long-distance telephone lines out of the hotel were broken.

At breakfast, the cafeteria counter displayed noodles with lard on them, vegetables still in the can, elephant-colored meatballs lying in liquid beneath a layer of congealed fat, some sort of internal organs in a cold, milky soup, and chunks of bony, deep-fried fish. I asked for fish, and the large, grandmotherly woman behind the counter took my money with one hand and with the other picked up a chunk of fish in her fingers and gave it to me. It was cold.

The second morning I asked for a chunk of fish again. The woman pointed to a tray of other fish, long and thin, which someone had just brought in. She leaned forward across the counter. "Take one of those instead," she advised, almost conspiratorially. "They're hot!"

Hearing my accent, she asked where I was from, and was so delighted to find an American that she impulsively offered to mend a hole in my sweater. Then she asked if she could bring to work tomorrow a ten-dollar bill that she had "bought from a Chinaman" at the hotel recently, so I could tell her if it was counterfeit.

An even more unexpected request came from another diner who overheard this conversation and then sat down next to me at a table where I was trying to take the bones out of my fish with a spoon (there were no knives or forks). "Could you," he asked mysteriously, leaning toward me, "by any chance sell me a size 44 woman's skirt?"

Magadan lies in a bowl rimmed by rocky hills; part of it spills over a final ridge, and then down to the harbor of Nagayevo Bay. In the center of town, Miron Markovich pointed out where Magadan's most dramatic try at building something new had come to an embarrassing halt. It was a multi-winged skyscraper that the authorities began putting up more than ten years ago, to be the grand future home of the regional Communist Party. But in the late 1980s construction came to a stop, clogged by shortages of supplies and money, and, perhaps, by the slowly collapsing Party's foreboding that its fancy new headquarters might end up being used by someone else.

Today the highest structure in the city is this ghostly, thirteen-story, gray concrete shell, surrounded by a ragged wooden fence. Some windows have glass in them, but even on the higher floors, most of this is broken. One upper window frame is blackened with smoke, as if high-rise squatters had built a campfire inside. Various people in Magadan want the building to become the home of a future joint Kolyma-Alaska university. But it is unclear if Alaska has any interest. Or who, for that matter, would pay for the cost of finishing the building, and of converting its hundreds of Party officials' offices into larger classrooms and lecture halls. "Now they haven't even got enough money to tear it down!" said Miron Markovich. Also, as with so many grand Soviet public works projects, there turns out to be a problem that nobody thought of beforehand. In time of flood, it is now said, the whole thirteen stories could slide into the nearby Magadanka River. And indeed, one day I saw several laborers on the riverbank, halfheartedly working on what appeared to be a dike.

"There are a thousand such unfinished buildings in this country," the director of the local TV station told me gloomily. "This one now attracts suicides, like that bridge of yours in San Francisco."

The skyscraper loomed above the city like a skeleton. As the cold, foggy afternoon turned into dusk, it was not hard to imagine these streets thronged with ghosts: long, gray columns of haggard men and women, just off the ships, on the first step of their journey from the harbor to the mines of the interior.

". . . the columns of prisoners marched around the clock day after day right through the town from the port to the disinfection center," writes Vassily Aksyonov, who lived here as a teenager, in his novel *The Burn*. ". . . In the predawn mist of the early morning, when Tolya walked from his suburb to school in the center, one after another those columns would come flowing toward him. Their characteristic sound was the shuffling of hundreds of shoe soles, the muffled, indistinguishable talk, the shouts of the guards, the growling of the dogs. . . . In time Tolya grew accustomed to the convicts and stopped noticing them, just as a pedestrian in the big city doesn't notice the traffic when he walks along the sidewalk."

Not far from the skeletal skyscraper is the site of a prisoners' dining hall, the scene of an episode Eugenia Ginzburg describes in her *Journey into the Whirlwind*:

After being arrested and interrogated in her home city of Kazan, in European Russia, Ginzburg endured the long journey to Kolyma. Newly arrived in Magadan, ill and exhausted, she managed to survive because she was assigned to work in this camp kitchen, where there were sometimes extra scraps of food. One day a group of emaciated prisoners arrived in Magadan, returning from the mines. One of them poked his head through the window into the kitchen where Ginzburg was working.

"Which of you's from Kazan?" he asked in a hoarse voice. I trembled at the wild thoughts that went through my head. Could my husband be among these dying men, or was it a message from one of my friends, and if so, who?

"One of the fellows here is from Kazan—he's on his last legs, he won't see the night through. He heard there was a Kazan woman working here, and he sent me to ask if there was a chance of his getting a piece of bread for his last meal. Could you spare some for him? You're so lucky to be where the food is."

Ginzburg asks who the man from Kazan is, and the other prisoner tells her:

"He's a Major Yelshin. He worked in the Kazan NKVD."

Major Yelshin had been Ginzburg's interrogator, the man who made sure she got a ten-year sentence instead of only five years. Trying to get her to confess, he had offered her food, when she had had none for days. ". . . could I forget those French rolls with slices of tender, pink, succulent ham . . . ?"

Finally she gives the other prisoner some bread for the dying major—and tells him to say whom it's from.

18
The
Inmost
Circle

A day or two after my arrival in Magadan, Miron Markovich and I rode the bus out to the airport to take the plane for Omsukchan—the base for our helicopter tour of old labor camp sites. I had been imagining that trip for months now, a voyage to the land of the dead.

The road that leads north out of Magadan to its airport is the beginning of the Kolyma Highway—the main route through the territory, dug and blasted out of the frozen wilderness, winding this way and that to touch various gold mines. Some of the spots where prisoners worked and died were named only by their distance from Magadan on the Highway: Kilometer 101 or Kilometer 237.

Miron Markovich was explaining all this above the roar of the bus engine. I found I was thinking of him as MM, for that was how I referred to him in the notes I took each day. From time to time as we bounced along, MM would take my pen and write or draw in my pocket notebook, to illustrate something he was saying. In between pointing out things we passed, MM was explaining what appeared to be his theory of human nature. His torrent of words began to reach beyond my vocabulary. Judging from the shaky diagram he drew in my notebook between bumps in the road, something—human society? Russia? The soul?—was divided into four quadrants. Were these the four points of the compass? Or the four humors of medieval physiology? Yes, that must be it, I thought, and tried to remember: choleric, melancholic ... But then MM began filling in points in the quadrants with various scribbles: "Hemingway," "Dostoevsky," "psychosis," "Tolstoy," "schizophrenia," and various other words which either I didn't know or which the potholes in the road made illegible. Luckily the bus got to the airport before I had to declare my own opinion on the matter.

Omsukchan is about an hour's flight northeast from Magadan; less than 1,000 miles farther and we would be in Alaska. On the sturdy, twin-engined propeller plane, many passengers were carrying food. From one woman's handbag projected the ends of two legs of lamb.

On landing, MM and I were met by a woman who edited the local newspaper, *The Omsukchan Worker*, and—his eye glued to a camera eyepiece—a man from the town's TV station. They said this was the first time any American, or any foreign journalist, had come to Omsukchan. Also at the airport to greet us was Alexei Lepiakin, the bear-hunting contractor who had arranged the helicopter for the next day. Lepiakin was an inspector from the local electric utility. His bear hunts for visitors—there had only been a few of them, he said—was a sideline.

Omsukchan's only hotel was a little inn with a dozen or so rooms, called the Dawn. Electric wires stood out in long ribs beneath the wallpaper. The hot water was shut off for repairs. This was now the third city on this trip where I had encountered the Great Annual Hot Water Shut-Off; like the sun traveling the wrong direction, it had followed me across the country from west to east.

The Dawn's restaurant was closed for repairs too, so Alexei Lepiakin's main concern was that MM and I get enough to eat. He insisted that we come to his apartment three times a day for meals, provided by his wife Natalya, who was as ample as he was thin. Every time we arrived, she had the table spread with an enormous meal: wine, vodka, chicken, peas, bread, rice, salmon caviar, whortleberry juice, and potatoes. Only the last two were produced locally. Omsukchan is the "potato zone," but the growing season is too short for other vegetables. They must have worked for days to get all this food together; hardly any of these items were on sale in the town's meagerly stocked grocery stores, and even a truckload of cabbages passing by had a soldier sitting on top, on guard. A bearskin was tacked to the wall above the Lepiakins' table. "Eat! Eat!" Natalya commanded, and we did.

I never did figure out how the Lepiakins' bear-hunting "cooperative" worked. Somehow everyone in town seemed to be in on sharing the excitement, and probably the rubles, of the rare foreign visitor. Most of the time we were driven around in a jeep that belonged to the newspaper; on its side, in large letters, was the word *Informatsiya* (Information). But sometimes a pleasant man named Sasha, a friend of the Lepiakins, drove us in a small truck borrowed from some workshop or factory. It was not clear to me whom tomorrow's helicopter belonged to, but everything was in order, Alexei Lepiakin assured me, because the head of the local Party committee had authorized the flight.

All these people were at first bewildered by MM. Puffing on his ever-present cigarette, he assumed command at all times. Besides his other quirks, I noticed, MM had two distinctive habits in speaking Russian.

The language has a second-person singular, *ty,* used with relatives, children, or close friends. But MM called everybody *ty:* town officials, bear-hunting contractors, helicopter pilots. At first I wondered if this was a habit from prison camp: in most languages that have the familiar form, prisoners, like soldiers, students, and workers, all use it with each other. But finally I decided that his was more the familiarity of the old family doctor, who assumes that everyone he meets he has, at some point, cured, comforted, operated on, or delivered.

Written Russian differs somewhat from the spoken language. Not for MM, however. His other linguistic peculiarity was a fondness for a whole class of words, participles, which are normally used almost only in writing. These are complex, thorny words, bristling nastily with prefixes, suffixes, and infixes, which, if you translated them literally, would read something like, "The having-stood-up Prince Andrei gazed upon the having-turned-toward-him Natasha . . ." Grammar books reassure the innocent newcomer that participles are generally not used in everyday speech. But MM was unaware of these promises. He spoke in participles. It gave him the sound of someone perpetually delivering a speech or reading a newspaper article aloud. And so he plowed through Omsukchan as he had through Magadan, emitting smoke and participles in all directions.

In the area around Omsukchan, 90 percent of the economy is still based on gold. The town had the extremely bare look of the Far North. Life is as hard here for plants as for people: when the roots of a bush or tree go down, they soon hit permafrost. Wood is so scarce that when Michael Solomon arrived as a prisoner in Omsukchan in 1949, he and other convicts were immediately sent out to the nearby mountain slopes, under guard, to dig up buried roots for firewood.

Whatever trees had once been here had long since gone up the chimney, and there were none to soften the rows of gray cinderblock apartment houses. Omsukchan's main street was made of solid concrete—in this climate, asphalt crumbles when moisture seeps into its tiny cracks and freezes. Many windows had three layers of glass against the cold. In front of the city government offices where we dropped in on the mayor, some shaggy horses from a local collective farm grazed on a patch of short-lived summer grass.

The afternoon of our first day, Alexei Lepiakin drove us in the jeep to a nearby hamlet, Galimyy, the home of a former labor camp guard who was said to be willing to talk. The road was of gray, powdery dirt. Along it drove big blunt-nosed Kamaz trucks from the gold mines, and sometimes a double-tracked *vezdyekhod,* or "go-everywhere" for traveling cross-country, that looked like a low, turretless tank. Each time a vehicle passed, we had to pull over and wait for its dense plume of gray dust to blow out of the way so we could see the road again.

Michael Solomon once worked in the mine at Galimyy. The underground tunnels were very narrow, he writes, and each miner had to carry his own crude lamp, a tin can of grease with a wick in it. These frequently went out, or, worse, ignited gases that had collected in the tunnels. Tree trunks were supposed to hold up the ceiling of the mine, preventing cave-ins, but these were often stolen by freezing prisoners desperate for firewood. Life above ground was little better: "To go to the latrines, situated a hundred yards from our dormitory, we had to hold tight to a rope or else be swept away by the howling blizzard."

Galimyy today is a small cluster of two- and three-story apartment buildings on a barren hillside. In one of them lived Vasily Udartsev, the former guard. His wife Alexandra greeted us at the door, a robust, talkative woman in a blue polka-dotted dress, with dyed red hair. Her husband would return in a moment, she assured us; he was finishing some work in their greenhouse out back.

Alexandra Udartseva herself first came to Galimyy as a school-teacher, she said, when the *gulag* was still operating full steam. At the beginning, "everything seemed exciting. They met us almost as if we were Martians, because before us there had been almost no women here. They brought us here at night: here was this high hill and those lights [guard tower searchlights] on both sides—it was like a Christmas tree!

"We saw convicts for the first time. 'Who are they?' we asked. 'It's no business of yours,' we were told. On Mondays, there would be a prisoners' band playing. There would be a soldier beside them, guarding them. They would play and we would dance."

It was at Galimyy that Udartseva met her future husband, who had already been a guard here for several years. He would join us very soon, she insisted, perhaps he'd gone over to help a neighbor repair a shed. He was just shy around strangers, a bit self-conscious about a scar on his face, he'd been in a motorcycle accident a while ago ... Meanwhile, "Have some tea! And try our jam of the North!"

What had her husband done, exactly, at the camp?

He was in charge of a squad of guards, Udartseva said. "He

told me awful things. Convicts escaped often. The law was that if you don't come back after warning shots are fired . . . So they killed them. I don't know how many. And they had to bring proof that they had killed them. One time they brought back a hand [for fingerprinting] as evidence. That particular one who was killed, I remember the name—Krasnov."

Then, pouring out with nervous animation, her narrative slipped seamlessly into the same kind of fantasy that had justified the horror of a half century ago. "Krasnov was a cannibal! He ate people. He would kill a man and eat him, just like that. Some people like that, bandits and marauders, were sentenced to two hundred years. His pals organized an escape for him. To Alaska! Many of them. They killed hunters to get their rifles, they killed soldiers and took their machine guns. They left traces of fires. They were caught at the airport in Magadan. The plane was already there, waiting to take them to Alaska. But the *Chekisti* arrested them all."

Udartseva excused herself to go outside for a moment, to try to persuade her husband to come in and talk to us. She came back shaking her head in embarrassed disappointment.

Finally I asked, did her husband (like prisoners I've talked to) still dream about that period of his life?

"Very often. At the beginning, he had terrible trouble sleeping. He suffered so much anxiety. Then sometimes he went on drinking binges. Finally he told me everything. And then he was afraid that he could be put on trial for that. . . ."

On the drive back to Omsukchan I found myself thinking about the story of Krasnov with the big gang of confederates and the escape plane ready to fly him to Alaska. (All this in a territory controlled by the NKVD, and in a country with no private planes.) If you could believe such things, you could just as easily believe that some helpless peasant was a Japanese spy or was conspiring to put the Tsar back on the throne. Udartseva's story also had some striking echoes of the witch hunts: this Krasnov was a cannibal (witches ate roasted children), left traces of mysterious fires in his wake (witches had their rendezvous with the Devil around midnight fires), and had a magical means to take the sky (witches rode broomsticks, of course, or airborne goats).

However implausible they were, you *had* to believe such things, in order to justify burning someone at the stake—or to bear to see your husband carrying a loose human hand. If you thought of the hand as having belonged to someone who was a human being like you, the sight would be intolerable. And so the innocent young schoolteacher had needed these myths to adapt to life in this strange new place lit up like a Christmas tree. Udartseva's fantasy, still intact after forty years, was a reminder of how so many people back then could have done the things that they did. If the people you do them to are criminal, dangerous, cannibalistic, magically empowered with networks of co-conspirators and airplanes, then aren't the very firmest of measures justified?

When we drove back to Omsukchan, we stopped on the outskirts of town to look at a mass grave that had been discovered there the previous year. "Anywhere where there was a camp, there is a grave," said MM. Some three hundred mass graves have been found so far in Kolyma. This one was next to an electric transformer station. Alexei Lepiakin pointed to the metal towers holding up new high-tension wires above us: "When they put those up, the bulldozers kept digging up bones."

Electricity hummed overhead; the long-drawn-out northern sunset shed a strange, black light on the shaded side of the rocky hills. Beneath our feet on the rough, stone-covered ground were ribs, leg bones, the base of a spine, forearms, bits of skull, a thigh bone still in its hip socket—all bleached white. MM picked up a lower vertebra and pointed out many curious tiny holes in it. "This is not from the wind or erosion," he said, "but from some kind of disease. Perhaps cancer or malnutrition. It would take a skilled pathologist to tell."

"The rock and the eternally frozen earth of the permafrost refused the corpses," writes Varlam Shalamov. "The rock had to be dynamited, broken open, prised apart. Digging graves involved the same procedure, the same instruments ... as digging for gold." Thomas Sgovio, the American, who survived several camps near Omsukchan, was once assigned to carrying frozen corpses to a mass grave. "We had to be careful lest we dropped them. An arm or a leg might snap off—like a dry twig."

Like the past they reflect, many of the country's newly discovered grave sites are not honored or even marked. A few days before, on my way to Kolyma, I had seen another mass grave in the Siberian city of Krasnoyarsk. It was in a corner of a cemetery; as the cemetery had expanded in the last year or two, workers digging new graves had suddenly begun finding masses of bones. On the day I was there, they still lay scattered and neglected on the ground. I saw mourners coming to lay flowers on regular, marked graves walk, without pausing, past dirt-yellowed skulls with bullet holes in them.

Here under the electric wires in Omsukchan, however, things turned out differently. Several months after I returned to the United States, the editor of the town newspaper, who had been with us that afternoon, sent me some photographs. A Russian Orthodox priest had come to town, and, before a solemn crowd in fur hats and heavy overcoats, had at last given these bones a formal burial beneath a wooden cross.

Omsukchan is in the "white nights" region, where in the summer it never gets completely dark. Late in the evening, under a ghostly gray sky, MM and I went to the Lepiakins' apartment for dinner. "Eat, eat!" said Natalya. Afterward, Alexei pulled out a map and we planned tomorrow's helicopter trip. At the prevailing rate of exchange, the helicopter was costing me only $30 an hour. No wonder Russia is going bankrupt.

After working out the itinerary, Alexei Lepiakin said, "Natalya and I will be coming with you, of course."

"Okay."

"And," he added, "we feel it's very important for our children to see these places where such terrible things happened in our country's history."

"Are you sure there's room?" I asked. "How many seats are there?"

"Then Oleg from the TV station, and Lyuba from the newspaper would like to come. And of course there's Sasha. He'll be bringing his boy."

"Wait a minute!" I said, "I'm not paying for a larger helicopter!"

Then Lepiakin explained: there is only one size helicopter in Kolyma. "The small ones are too fragile." At less than the rate for a New York taxi, I was getting a twin-jet helicopter with a crew of four—and room for twenty-two passengers.

The huge, bright orange machine is waiting for us on the airport tarmac the next morning. And Lepiakin is right: every other helicopter at the Omsukchan airport, or that I see anywhere else in Kolyma, is the same size. Because few roads cross Kolyma's rocky mountain ridges, helicopters are the workhorses of the territory, carrying everything from mail for remote reindeer farms to parties of geologists hunting for gold.

Here I am at last, on my way to the Dachaus and Buchenwalds of the North, the long-envisaged climax of my solemn pilgrimage to the dark heart of the Soviet past ... but the journey is not following the script. This is a party. The three children on board are excited. In addition to the crew, there are seven adults, and to feed us all Natalya is carrying an enormous picnic basket. She begins passing out sandwiches and cookies before we are airborne. Oleg from the TV station is a great fan of Hollywood gangster movies, and asks me questions like, "When the American police arrest somebody, do they *really* throw him up against the side of a car and feel all his pockets for weapons?" And I, for the first time in my life, get to sit in a helicopter cockpit, on a fold-down seat between the two pilots.

Automatically, I reach for the seat belt. But the pilots have a macho contempt for such things. "*Ne nado!*" (You don't have to), they say, the same phrase I've heard a dozen times from Russian cab drivers. In any case, there's no buckle.

In a moment, the dirt airstrip at Omsukchan is out of sight behind us, and we are skimming along several hundred feet off the ground. We lift exhilaratingly over the crests of ridges, and roar through the gaps between *sopki*—the bare, rocky hills several thousand feet high that form the basic terrain of Kolyma. Many *sopki* are crisscrossed with tracks left by the jeeps and tracked "go-everywheres" of the geologists. It makes them look like giant Arctic versions of American hills scarred by dirt-bike trails. The valleys between *sopki* have fir trees, but low and sparse: less than 400 miles north of Omsukchan no trees will grow. On the horizon is a wall of large, snow-covered mountains. The co-pilot repeatedly claps his hands, trying to kill the mosquitoes flitting around the cockpit.

Above the shriek of the engines, the pilot calls out, "A bear!" The helicopter dives down in a steep spiral toward the top of one *sopka,* breathtakingly close. An enormous brown bear is grazing contentedly on the rocky summit. His breakfast interrupted by the clatter of the rotor, he runs frantically down the slope, clumsy hindquarters sticking high in the air.

We cross the Kolyma Highway, a ribbon of gray dirt; along it are dotted mile-long dust plumes trailed by slow-moving trucks. On this road in winter, you never turn off your truck engine; it won't start again. From the cavernous, echoing interior of the helicopter, MM roars out that they built small camps every 15 kilometers along the highway to hold prisoners in the early days, when they were marched into the interior on foot. We swoop past abandoned prospectors' huts on the hillsides, and past a radio tower, where a man runs out of a shed to look skyward and wave. Updrafts along the sides of the *sopki* waft the helicopter higher in the air. A winding river below is muddied to a dirty green. The pilot shouts that this means they're mining gold upstream.

We fly from east to west for some two hours, although this by no means takes us all the way across Kolyma, and from south to north the territory is much longer still. Its very size is a token of the power of the NKVD. For Kolyma has a strange distinction. From 1931 until after Stalin's death, the entire region was administered outside the ordinary channels of Soviet government. Elsewhere in the country the *gulag* was, in Solzhenitsyn's famous phrase, an archipelago, most of its camps scattered among normally governed towns and collective farms. But in Kolyma regular departments and agencies of the Soviet government did not operate. Instead, all inhabitants up and down the 1,000-mile length of the region, both prisoners and the thin scattering of "free workers," were directly under the authority of the secret police. Kolyma was, literally, a police state.

Beneath the vibrating helicopter the *sopki* gradually turn higher and more forbidding, their steep, gravelly slopes more densely streaked with snow. Then suddenly in a long, rocky cleft below us is our destination: piles of gravel mine tailings, and the roofless stone ruins of many buildings, strung out for maybe a mile through a narrow valley. The helicopter drops down until we are only two or three feet off the ground. A crewman jumps out to make sure we won't be landing on earth turned boggy by patches of melting snow, then motions palms-downward to signal the pilots that it's okay to land.

We are at Butugychag. This camp complex once held more than twenty-five thousand inmates, according to the poet Anatoly Zhigulin, who was a prisoner here in the early 1950s; still more were in four satellite camps nearby. One of those, a tin mine, whose ruins I can make out on top of a *sopka,* was exposed to the high-altitude winds, and was without water, which prisoners had to carry up from below.

We think of concentration camp barracks as being made of wood, and elsewhere in the Soviet Union they usually were. But in Kolyma this was true only of the lower-lying camps near the forests. Zhigulin, like most Kolyma prisoners, lived in a communal, double-walled tent. As insulation against the ferocious cold, prisoners piled moss against the outsides of these tents, and between the two layers of cloth. On Zhigulin's first night here at Butugychag, a thief slit the tent wall and stole his coat. He survived only because one or another prisoner coming off the night shift loaned him a coat each day to wear to the mines.

The former women's camp in the Butugychag complex is eight miles away, around a bend in the rugged valley. Squads of prisoners came from it to the main camp each day to pick up food and carry it back in knapsacks, "human ants carrying their heavy loads up the rugged mountain," writes Michael Solomon, who also passed through Butugychag.

> ... Among the women assigned to this heavy task I recognized a famous young actress ... Nadya Milionushkina, winner of the all-Soviet Union contest for drama in 1948. Tall and slender, with thick auburn hair down to her shoulders, beautiful green eyes, and a milky complexion, Nadya looked like a queen even in the shoddy camp clothes she had to wear ... not only did ... prisoners fall in love with her, but even ... *politruks* [NKVD political commissars]. She pretended to accept the courtship of the *politruks* until these unsuspecting men invited her to their offices. Then when they tried to make love to her, she would begin to scream, smash windows, bang on doors, and create such a scandalous din that all sorts of curious people immediately arrived on the scene.... The result was that the *politruks* had to be transferred somewhere else because they had lost face in front of the prisoners. I asked Milionushkina why she had continued to play such a dangerous game. Her answer was that she wanted to see them disgraced or, more hopefully, shot for misbehavior.... Lifting the heavy bag of food on her shoulders, she smiled at me before taking off and said: "Believe me, there is nothing more pleasant than to see *Chekisti* dying to make love to you and then being shot as traitors to their motherland."

A special hazard faced prisoners at Butugychag—radiation. The mines here produced both coal and uranium. So much radiation has been measured at this spot in recent years that we've been warned not to stay long, and not under any conditions to enter the old mine tunnels whose mouths we can see on the slope above us. Couldn't we just take a quick look inside? I ask. "Don't try it," says MM, who has radiation statistics among his storehouse of *gulag*-related information. He reels off numbers about alpha levels. "Your descendants will become genetic mutants. Just as we're already psychological mutants here in the Soviet Union. . . ."

Many prisoners at Butugychag never lived long enough to suffer from radiation; they collapsed of exhaustion while working, or were killed in the frequent mining accidents. When this happened, sometimes their bodies were simply thrown down unused mine shafts, so they would not have to be carried back down the mountain and buried.

As we walk through the ruins, piles of snow, grimy with rock dust, still lie on the ground even though it is summer. We are less than 300 miles from the northern hemisphere's Pole of Cold. In his gold-mining camp, writes former prisoner Vladimir Petrov, "Soup still warm when received . . . would become covered with ice during the period of time one man would wait for a spoon from another who had finished with one. This probably explained why the majority of men preferred to eat without spoons. . . ."

It feels strange to be standing at last in the inmost circle of the prisoners' inferno, one of the worst camps in the worst place, Kolyma. Rocky, treeless, frozen, radioactive: few labor camps came worse than this. But I find that the image that was in my head is different from the one in front of my eyes. Despite everything I had read, the Kolyma camp of my imagining had somehow become a sort of picture-book German POW camp of World War II movies, closed in by barbed wire, on whose other side lies Freedom. Rotting fence posts and rusted tangles of barbed wire do indeed now lie all about us on the ground, but on the other side of this wire there are no symbols of civilian life, no forest full of beckoning hiding places, no signs of freedom. Instead, a prisoner who escaped would have, in most directions, an arduous climb of up to several thousand feet—to his workplace, the mine tunnels.

Just as the camp lay within an imprisoning landscape, so did it hold a further prison within it, a jail within a jail. I step inside this building at one end of the Butugychag camp. Its thick stone walls are built with extra care; it is the only one of all the stone ruins I can see here where real cement, not mud plaster, was used as mortar. A dozen or so small rooms open off a corridor. Their windows are barred—vertically and horizontally. And at all the points where these two sets of bars cross, they are fastened together with thick, chainlike links. The cells look as if they were built not to hold starving unarmed convicts, but gangs of metalworkers imprisoned with all their equipment.

The significant thing about each cell, MM points out, is not how many people it could hold, but how far away it is from a rusted tangle of metal at one end of the corridor—the remnants of the building's stove. "Some of these were luxury cells," MM says, "closer to the heat. Then there were cells where the heat would never reach. Anyone who said something nasty to the warden could be stuck in that cell, there, at the cold end of the building."

All camps had their "internal prisons" like this. A Ukrainian prisoner writes of one at a camp about 100 miles north of here:

In summer each day at 5 p.m. a fire pump was used to force water into the jail until it covered the floor to the depth of 5–8 cm. Those prisoners who had not performed 50% of their norm were driven in here to sleep. They passed the whole night on their feet for they could not lie down in the cold water and in the morning they were sent into the mine to work. . . .

. . . In December, 1938, more than half of the prisoners in the jail were badly frostbitten and few lived until the summer of 1939.

How much the harshness of life at a place like this must have been added to by its total isolation. The rest of the world did not know that the prisoners here existed. Or that such a place existed. Today, researchers believe, there are some Kolyma camps whose names we still do not know, because no prisoners returned from them. Looking up at the bare *sopki* of gray-black rock, it is easy to see why the same image came to so many Kolyma prisoners: the dark side of the moon.

Even in remote corners of the *gulag* like this, men and women sometimes found people they had known in another life. I heard about several such encounters from Nina Savoyeva, a retired doctor I talked to in Moscow. With the jet-black hair and eyes of her native Ossetia, in the Caucasus, Savoyeva was a legendary figure who shows up in several prisoners' memoirs. She was famous for her explosive temper, and for boldly trying to save the lives of many prisoners by claiming that they were too sick to work, or that she urgently needed their services as hospital orderlies.

Savoyeva was assigned straight out of medical school to a clinic at a Kolyma mine. She was in her office one day when a sick prisoner was brought in, badly frostbitten. "His face was invisible, it was covered with frost. The nurses helped, and I asked him to take his clothes off. He didn't take off his hat. I said, 'Take off your hat.' "

At once Savoyeva recognized him: it was her former chief surgery professor. He had disappeared suddenly one day in 1937—"transferred to the Leningrad medical school," the students had been told.

Later, Savoyeva was put in charge of a prison hospital about 100 miles north of Butugychag. After prisoners saw her making rounds the first day, someone delivered a gift to her from a patient: "A cigarette box with a picture of a Caucasian horse and rider; they are flying, with Mount Elbrus in the background. And on this box was a letter this patient had written me, saying that he had learned from the nurse that I was from the same village as he. He wrote me where his parents were, and that he was dying, and that he was innocent—a terrible and touching letter on the side of the cigarette box. He said, 'I'm giving you the most valuable thing I have, this box.'

"He didn't remember that I had once gone dancing with him, when I was a girl. . . ."

The next camp we land at, Urchan, is again bordered by these bare, rocky hills. Again barbed wire, again an internal prison with barred windows. The hills that box in this camp are so steep that the helicopter pilots cannot reach their ground controller on the radio.

The barracks are wooden this time. The floor has rotted from years of melting snow blown in through cracks and broken windows, and rusty nails project from exposed beams. A dead small animal lies in one corner. Like archeologists, we poke through the rubble for clues to life here. One is an iron tool that looks like an oversized, homemade spatula. Perhaps it was used, MM thinks, by prisoners who were put to work dressing the skin of reindeer or bear shot by a camp officer. We also find an ancient shoe. Years of melting snow have shrunken the leather upper part so much that the shoe has curled almost into a tight circle, absurdly small, like shoes for the bound feet of women in China.

Once in the air again, we fly north, making a stop to refuel at a small airport. Rivers and lakes beneath us are sometimes still partly frozen over with bluish ice. One body of water is called Lake Jack London. I am still up in the cockpit, between the two pilots. They are eager to talk and undaunted by the noise of the twin jet engines a few feet above our heads. Their questions come in shouts:

"HOW MUCH MONEY DO HELICOPTER PILOTS IN AMERICA MAKE?"

"HOW MANY HOURS A MONTH DO THEY HAVE TO FLY?"

"WOULD THERE BE WORK FOR US IF WE CAME?"

Because of the din, I have to scream my response into the ear of the co-pilot, who then relays it over an intercom to the pilot a few feet away. They chat, changing dollars into rubles and shaking their heads in awe. Then the familiar question:

"AND WHY ARE YOU SO INTERESTED IN . . . ALL THIS?"

I shout about books I've read, mentioning writers who were Kolyma prisoners, like Varlam Shalamov and Eugenia Ginzburg.

"GINZBURG. OF COURSE!" shouts the co-pilot. He makes a thumbs-up sign. "SHE WROTE BETTER ABOUT ALL THAT THAN ANY MAN." He and the pilot talk on the intercom, and suddenly the helicopter veers to the left. We are, it turns out, very near Elgen, the women's camp where Ginzburg was held. Ten minutes later, the camp site is below us, in a wide, flat valley with stubby fir trees.

It was at Elgen—the word means "Dead" in the Yakut language—that Ginzburg was sent into the forests to work, and was introduced to the quota system: cut eight cubic meters of wood a day or else your food ration is slashed. "You'll only eat as much as you'll earn." Unused to physical labor, weakened by the long journey, she couldn't make her quota; her bread ration was reduced. It was also at Elgen that she learned that the first frost of the Kolyma winter comes in August. One morning she awoke in panic, her neck paralyzed, she thought—only to discover that her hair, damp from snow drifting through cracks in the barracks wall, had frozen to her bed.

> . . . salvation from death in the Elgen forests came to me from cran-
> berries, sour bitter northern berries . . . coaxed out of their hiding
> place by the timid Kolyma spring, after their ten months' sleep under
> the snow. It was already May when, as I was crouching close to the
> ground in order to cut the branches off a felled larch tree, I noticed
> in the thawed patch near the stump that miracle of nature—a sprig
> with five or six berries on it, of a red so deep that they looked almost
> black, and so tender that it broke one's heart to look at them. . . . If
> you tried to pick them, they burst in your fingers; but you could lie
> on the ground and suck them off the branch with your dried,
> chapped lips, crushing each one separately against your palate and sa-
> voring its flavor. . . . From then on we went into the forest not in de-
> spair but in hope.

Only a few minutes beyond Elgen, we pass the end of the forest, and swoop down to our last camp of the day, Seymchansky Canyon. The ruins of at least thirty or forty wooden buildings are spread out over a valley floor. On the slope above, up a stairway, are the ramshackle remains of a big gold ore-processing plant. Poles for an ore-carrying cable railway march down a hillside across the valley, like a ski lift. The buildings are surrounded by miles of barbed-wire fencing and the ruins of watchtowers, known as *ptichniki,* or birdhouses.

The changing of the guard at such posts always followed a certain ritual, designed to emphasize that the prisoners were dangerous fanatics, to be shot immediately at any attempt to escape. "... in the evening silence, I could hear the guards handing over the watch in the towers around the camp," writes one former prisoner. "They had their own prayer: 'Sentry number forty-one. Post number three. For the defense of the Soviet Union. Guarding terrorists, spies, murderers, and enemies of the people. Sentry forty-one delivers the post.' "

Roughly two thirds of Kolyma's prisoners worked in gold mines like this camp at Seymchansky Canyon. The surface of the ground was frozen for most of the year, and miners had to break up the rock-hard earth with picks and shovels. Underground, the digging was slightly easier, because pipes took steam into the mine tunnels to thaw the ore. Despite its welcome warmth, the steam brought other problems. Thawed-out chunks of rock dropped out of the mine ceilings. And the steam was soaked up by the prisoners' crude and inadequate boots. "They absorb moisture with incredible speed, especially when the sacks of which they are made were used for bagging salt," writes Elinor Lipper.

Worst of all was the sudden temperature change from the mine tunnels to the outside. Vladimir Petrov, once a prisoner at At-Uryakh, almost directly beneath our morning's helicopter route, writes: "From the steamy, damp atmosphere ... the perspiring wheelbarrow-pushers slipped through the opening, which was covered by an old blanket, rolling out their wheelbarrows into the piercing 50 below zero frost. The time limit in this work was, at the most, one month, after which either pneumonia or meningitis dispatched the worker into the next world."

Oleg the TV man is following us about, and MM periodically turns to the camera with explanations. "A typical camp toilet!" he booms, as we peer inside a small log-walled shed. There is room for four or five people, sitting over a ditch, on two split logs with a six-inch-wide space between them.

"This was one of the largest camps," MM continues as we walk through Seymchansky Canyon. There are so many camp buildings that the dirt tracks between them were named. "Blue Street," says one battered sign, and one building has a number on it: 42. Who got mail at 42 Blue St., Camp Seymchansky Canyon, Kolyma, USSR? An administrator of some sort, it seems; maybe even the camp commander: In the ruins of the house are some benches, a battered desk, an empty safe.

The Seymchansky Canyon camp is on a creek, and takes its name from a town downstream, out of sight around a nearby mountain shoulder, Seymchan. This remote valley is one part of Kolyma that I am not the first American to visit. Another preceded me, by nearly half a century: the Vice President of the United States.

19

The Power
of Facing
Unpleasant Facts

In the summer of 1944, the wartime alliance between the United States and the Soviet Union was at its high point. The Soviets were delighted that the Western Allies were opening the long-hoped-for second front against the Nazis in France. The Americans were grateful that the Red Army now had the Germans reeling in retreat. When President Franklin D. Roosevelt sent Vice President Henry Wallace on a visit to Russia, there were feasts and toasts to victory at every stop.

Despite their strenuous efforts to keep the extent of the *gulag* secret from the rest of the world, the Soviets were eager for Wallace and his party to make a stopover in Kolyma. The Lend-Lease program for sending supplies to Russia was then in full swing. To keep the flood of American food and arms flowing, the Soviets were keen to show that they could eventually pay for it. They wanted to show the Vice President some gold mines.

Wallace spent three days in Kolyma. He visited Magadan, toured several mines not far from Butugychag, and spent the night at Seymchan, the town around the corner of the mountain from the massive labor camp of Seymchansky Canyon. It was probably the only time that a delegation of high-ranking Western government officials visited anywhere in the heart of the *gulag*. In the long, curious history of naive foreign visitors to the Soviet Union, this is the most spectacular example of non-seeing.

The Soviets were skilled in presenting what they wanted foreigners to see. In fact, the practice of taking a political delegation of visitors from abroad on a closely chaperoned tour of selected sites is a Soviet invention, dating from the early 1920s. The Soviets worked hard to give the Wallace group the impression that they were visiting a cheerful Russian Klondike full of happy gold miners—and they were wildly successful.

Watchtowers along the Kolyma Highway outside Magadan were pulled down. The hungry, ill-clad prisoners normally at work under armed guard paving the city's streets and constructing its buildings received, probably for the first and last time in their terms, a three-day holiday, so they could stay in their barracks and out of sight. NKVD clerks and secretaries were rushed out to a pig farm Wallace was to visit, to play the roles of farmhands. A photograph from the time shows Vice President Wallace shaking hands with a group of Kolyma "miners": one has his cap off and his head is shaved, indicating that he is a prisoner; but all look well fed, an almost certain sign that they are common criminal trusties.

"The Kolyma gold miners are big, husky young men who came out to the Far East from European Russia," Wallace wrote in his book *Soviet Asia Mission*, which appeared soon after the trip. He also waxed enthusiastic about the eight-hour Soviet work day and the high wages of Kolyma workers. Another U.S. official traveling with Wallace, the noted China expert Owen Lattimore, wrote an upbeat account of the trip for the *National Geographic*. Both men praised the agency that ran Kolyma, Dalstroy—the acronym of the Far Northern Construction Trust. Neither appeared to know that Dalstroy was a branch of the NKVD.

Some of the NKVD officials they met evidently took off their uniforms for the three days the Wallace party was in Kolyma. Lattimore blithely describes Dalstroy as "a combination Hudson's Bay Company and TVA." He goes on to say, "There has probably never been a more orderly phase of pioneering than the opening up of Russia's Far North . . . the over-all vision is one of communities which are well rounded and self-supporting. . . . It was interesting to find, instead of the sin, gin and brawling of an old-time gold rush, extensive greenhouses growing tomatoes, cucumbers, and even melons, to make sure that the hardy miners got enough vitamins."

The American delegation went to the ballet in Magadan, where prisoners danced for NKVD officials. The Kolyma NKVD boss at the time was Ivan Nikishov, whose wife was head of Maglag, the camp district of Magadan. Both were notorious for their cruelty and for their love of luxury—they had a lavish private hunting lodge and nature preserve. Lattimore writes of Nikishov that "both he and his wife have a trained and sensitive interest in art and music and also a deep sense of civic responsibility."

Wallace visited Karaganda on this same trip, and managed to come away with good impressions there, too. The Vice President's Soviet hosts must have been frantic about what he might see, and though Wallace never seemed to realize this, in his book on the trip he inadvertently drops two amusing clues to the Soviets' anxiety: "On the first day at Seymchan, I went out for my usual morning walk. As I started out one of the Russians promptly joined me for a hike in the evergreen forest. . . . Tramping over the permafrost in this way helped us to get acquainted. . . . Thereafter I never lacked a companion when taking my morning exercise." On this walk, Wallace clearly never got within sight of the camp at Seymchansky Canyon. One afternoon in Magadan, Wallace also went for an impulsive walk, he reports, only to find a Soviet officer running after him, shouting, "Dinner is ready."

To his credit, Wallace later realized he had been taken in. In 1948, four years after Wallace and Lattimore's visit to Kolyma, Vladimir Petrov, one of the first Kolyma survivors to reach the United States, published some magazine articles about his ordeal. Wallace saw these, sought out Petrov, questioned him about his experiences, and the two men became friends. Wallace himself then publicly apologized for how he had let himself be fooled in Kolyma. Lattimore never did. When criticized by *The New States-*

man on this score several decades later, he somewhat flippantly defended himself in a letter to the editor. But the striking thing is how easily both men were taken in. Wallace spoke some Russian; Lattimore was a distinguished China scholar, as was John Carter Vincent, another member of the party.

Many prisoners themselves wondered how a group of otherwise intelligent, knowledgeable people could spend three days in Kolyma and not see that the whole place was a prison. The Wallace trip became something of a legend among them. Americans asked the same question nearly a decade after the visit, when Wallace and Lattimore became targets of brutal attacks by Senator Joseph McCarthy and his allies. Lattimore was hounded out of his job, and, for a time, out of the country. The accounts each man had written of their visit to Kolyma were cited against them, and they were both viciously attacked as Communist dupes.

The McCarthyites ignored, however, the fact that the U.S. military officers traveling with Wallace and Lattimore had also seen nothing amiss in Kolyma. Which leads to a larger point: if you look at the whole array of Westerners who visited the Soviet Union during its first quarter century and who either did or did not see the executions, the prisons, and the famines, the division between those who saw clearly and those who denied everything did not always break down along ideological lines.

In the 1930s, two of the worst deniers of all were major figures of the establishment: U.S. Ambassador Joseph Davies, a staunch capitalist, and *New York Times* correspondent Walter Duranty. Both sent home all-is-well messages from Moscow at the height of the Great Purge. Davies thought the arrests "cleansed the country" and "rid it of treason." Duranty agreed, and totally whitewashed the great collectivization famine as well. The denials of both men were all the more remarkable because they were not two-week visitors on a guided tour; they were living in Russia and knew some of the Purge victims personally.

Although many left-wing intellectuals brought back appallingly cheerful accounts of short trips to Russia during the 1920s and 1930s, a brave handful did not. André Gide, whose sharply critical 1936 book, *Le Retour de l'U.R.S.S.*, was the biggest literary bombshell from any of these visitors, was then a member of the French Communist Party. And two of the very first Westerners to speak out against Soviet injustice were the arch rebels Bertrand Russell and Emma Goldman. Both visited the infant Soviet state soon after the Revolution. Each met Lenin, and was struck by his cruelty; each writes of hearing shots in the night in Petrograd—the sound of political prisoners being executed. Both visitors upset their Soviet hosts by asking for various things not on the program. Goldman demanded to see a prison.

Other travelers, however, brought home the rosy image with which they went. They had erected the Potemkin villages in their heads before departure, and they projected this picture onto the Russia they arrived in. Any traveler determined in advance to find a promised land will usually find it.

Often the picture-in-the-head had little to do with a formal belief in communism. One of the most vocal British fellow travelers was the Dean of Canterbury, whose visit to Russia convinced him that the Soviets had "recovered the core of real belief in God." Others embraced Russia in part to outrage the stuffed shirts at home: in 1931, George Bernard Shaw threw away a parcel of food before his train crossed the border into the Soviet Union, because he didn't believe the British press, and, he cheerfully told people, he knew there would be plenty of food in Russia. (For him, there was.) Shaw's picture-in-the-head of Order and Progress was so vivid that, like another famous Soviet admirer, Lincoln Steffens, he for a time enthusiastically projected it onto Mussolini's Italy.

When Henry Wallace and Owen Lattimore toured this Kolyma valley, they, too, carried a picture in their heads that blocked their view of the concentration camps: the Soviet Union was our brave wartime ally. Wallace and Lattimore went to Russia on an official mission; part of their job was to bring home good news. In their non-seeing, they reflected the mood of the wartime American elite. The year before, Henry Luce's *Life* magazine had published a special issue about Russia, filled with photos of graceful ballerinas, gallant soldiers, and cheerful tractor drivers. If anything,

Life went even farther overboard than Wallace and Lattimore, calling Lenin "perhaps the greatest man of modern times," and the NKVD "a national police similar to the F.B.I."

And the clear-seers? Although not on official missions, Bertrand Russell and Emma Goldman had every possible motive to find hope and good news in Russia. Both were deeply committed to the progressive movements of the day. Both would have loved to see a land where workers had power, women could vote, and socialism worked. Both did not hesitate to dream, in the best sense of the word, in Utopian terms. Russell not only advocated all sorts of political and economic reforms, but proposed radically new approaches to education and marriage as well. Goldman had just been deported from the United States because of her politics. Yet when they arrived in Russia, they indulged in no wishful thinking.

Both Russell and Goldman had a long history of saying what they believed, whether it was popular or not. Only a few years before seeing Revolutionary Russia, Russell had gone to jail in England, and Goldman in the United States, for their principled opposition to World War I. No social pressure to conform is stronger than jingoism in wartime, and seldom was jingoism stronger than during that war. Russell and Goldman were both reviled as traitors who stabbed their countries in the back in an hour of need. People strong enough to survive that could reach their own judgments about Russia, without help from tour guides or the political fashions of the moment. Theirs was the courage of perception, what Orwell calls the "power of facing unpleasant facts."

In the ruins of the Seymchansky Canyon camp it is now evening, and that strange black light lies over the Kolyma hills. MM and I, with our odd assortment of traveling companions, climb back into the helicopter for the last leg of today's flight. We wind our way out of the valley, past the town where Wallace stayed, between the tops of the rocky *sopki,* back toward Omsukchan. And even though most don't appear on the maps of known Kolyma camps, on this final part of our route it seems as if every valley holds its old prison camp. We fly over more than half a dozen as the helicopter turns this way and that: collapsed remnants of barracks and watchtowers and gold-mining works, the wood blackened by decades of snow and sun and rain. The ruins lie around every bend, beyond every ridge, as if a giant hand has angrily thrown them down from the sky every few miles across the barren moonscape.

The next morning I wake up in my room in the tiny inn at Omsukchan and notice that it has a small TV set. What would they be broadcasting up here at the top of the world: Reindeer prices? Bear meat recipes? The permafrost report? I turn it on.

I have forgotten, however, that most TV in Russia is national, and that what people want to watch now are programs, any programs, from abroad. On the snowy, blurry black-and-white screen, I recognize Frances Lear, whom I once sat next to at a dinner party on what seems from here like another planet. She is chatting about how easy it was to start *Lear's* magazine. Beneath the Russian voiceover I can hear her saying to an interviewer: "It's easy, Bob. If you've got a good idea, and you've got some good people to work with you, and you've got faith in your idea, then *anybody* can do it . . ." She turns out to be one of a string of American multimillionaires briskly telling someone from Cable News Network how they made their multi-millions. Here, in the very graveyard of the old Utopia, they are unveiling the icons of the new. Poor, poor Russia: will this new set of illusions be any more useful than the last?

We are due to fly back to Magadan today, but in the morning Alexei Lepiakin arrives with bad news:

"There's a cyclone coming. The airport is closed. Natalya says you must come and have something to eat."

"Could we go back to Magadan by car?" I ask.

"Perhaps. But it takes twelve hours. The good part is that the rain will keep the dust down on the road. The bad part is that the cyclone may wash out some of the bridges."

We decide to stay in Omsukchan and eat.

At my hotel in Magadan, the long-distance telephone lines had not been working. Oddly, here in even more remote Omsukchan, beset by a cyclone, they are fine. From the Lepiakins' apartment, I call my wife in Moscow.

"The press department at the Foreign Ministry called," she says. "They're upset that you haven't given them your flight numbers." I explain about the broken phone lines and the storm. And, I wonder, do they even have flight numbers out here? Alexei Lepiakin tells us that once the cyclone blows past, Magadan will simply send all available planes to Omsukchan to pick up stranded people. As wind and rain howl past the Lepiakins' windows, I feel like an astronaut, stuck in orbit, while somebody radios up from Earth to say that he missed a dentist appointment.

MM also calls his wife, back in Magadan, to report the delay. "She says to ask you if I'm getting on your nerves," he tells me cheerfully, his hand over the mouthpiece. "She says to ask if I'm talking too much!"

That afternoon, I sit MM down in front of my tape recorder and ask him to tell me the story of his own arrest and imprisonment. He was jailed as a young medical student, and spent three years in the *gulag*. Five years after his release, he came to work in Magadan. One of his first jobs as a psychiatrist was to help people recover their memories—an appropriate prelude to his work with Memorial. The most difficult cases he treated were "unknown people with unknown identities," labor camp veterans so traumatized by the experience that they suffered amnesia.

I ask if he treated any former guards or camp officials.

"Many. For me they were no different. For a Westerner, it may be hard to understand that between a guard and a regular person there could be no difference. But how can we draw a line between a guard and, say, an ordinary petty journalist, who in essence is also a kind of guard, but behind an editor's desk? He's even worse than the guard! The guard is out there in the open, with his gun. The other fellow would sit and talk with you. But at the same time, he's your guard, because the things you say in his presence, he'll later use to inform on you. What's the difference?

"On Gorky Street [in Moscow], one man met his former inter-rogator and rushed at him and slugged him. That's one type of such a meeting. But in 1956, when I came home from prison and was on my way to the house where my wife and mother-in-law lived, where I had been arrested, I ran into my former interrogator, Bubishev. We rushed into each other's arms and hugged.

"I can't explain it. But I remember I was then seized with ter-ror. What am I *doing*? Who am I hugging? In a jiffy, he took me around the corner of the movie theater where we had run into each other. And he said, 'Miron, I've survived too. But don't believe that everything is over. Keep quiet. Don't talk too much.'

"For that moment, we were happy to see each other. He had thought I was doomed. He hadn't tortured and humiliated me, he had done his job, and was a kind person by nature—I knew that. Toward me he had been a human being."

MM also talks about his father, who fought in the Russian Civil War, and later held a high post in the construction industry. A war wound made him almost deaf, yet he heard the warning notes in the air more clearly than anyone around him. One day in the late 1930s, he came home and told his family, "I think they're coming for me." He went into hiding, and for some years changed his ad-dress and place of work every few months. He survived. In nearly six months of talking with people about mass murder, this is the first time anybody has told me about a person who tried to escape it. I have read of a few other such cases. The NKVD was far better at inspiring terror than at ordinary detective work, and those who tried to avoid arrest by lying low and moving from place to place had a good chance of success. However, people rarely tried; despite the mass arrests, almost everybody believed, *It won't happen to me.* People deny bad news because it implies worse news: If I'm about to be arrested, that would mean the whole system has gone mad.

And today? If there is a greenhouse effect, a depletion of the ozone, a shrinking of ocean fish stocks and an expansion of deserts, a steadily widening gap between the world's rich and poor, then that, too, means that the whole system has gone mad. But the anal-ogy is imperfect. For we are free to read and write and talk endlessly about the greenhouse effect or the ozone layer or all the other problems, hence we do not feel the intense fear produced by the NKVD's knock on the door. That very lack of urgency is our form of denial, as foolhardy as the denials of the fellow travellers. For the knock, from all these things, will come.

20
Port
of Arrival

The next day the cyclone has blown past, and MM and I fly
from Omsukchan back to Magadan.

I have two days left before I have to return to Moscow. If I'm
out of MM's sight, at a meeting I've arranged on my own, he telephones in mid-interview, just to make sure I've gotten there. Many
Russians are hazy about appointments and deadlines, but not MM.
If he picks me up at the hotel, he always arrives precisely fifteen
minutes early. When I accuse him of doing this deliberately, he
cheerfully admits it, and says that it's "to make sure everything is
in order." When I go to his apartment for dinner one evening,
insisting that I can find my own way, he is out on the staircase
landing, smoking, waiting for me, pacing like a restless sentry.

A somber note sounds on one of these last days in fog-
shrouded Magadan. I ask if Memorial can find me some ex-
prisoners who had been in any of the camps we had visited in the
helicopter. No problem, MM says, "we have half a dozen such people living right here. We've got their names in our card file." He
and a colleague call several men and women, and send messages to
those who have no phone. But of the six people, not one is willing
to talk.

It is a sobering reminder of something I've dimly known in
theory all along, but which I keep forgetting. All the hundreds of

former prisoners who have published *gulag* memoirs, all the dozens of them like MM who have so willingly told me their stories, are the healthy exceptions. For most survivors, the pain is too deep to share, perhaps even with those close to them, much less with a stranger. For them, the wounds are still unhealed.

One person in Magadan who is eager to talk, however, is a writer and editor named Alexander Biryukov. Tall and earnest, he wears a brown leather jacket, a necktie, and tinted glasses. There is something driven about him; it is as if he has been waiting to speak for many years, and now that he has the floor at last, he has absolutely no time to lose. During the long, gray Brezhnev years, he "bumped his head," as the phrase goes, almost losing his job as an editor because he published a few lines of poetry that mentioned the *gulag*. He quotes one of the offending lines: "Between them lies the night, Kolyma, stars and losses . . ."

During the last few years everything has changed, and in recent months Biryukov has even been allowed to look at some files in government archives here. He is compiling several books by and about the many writers who perished in Kolyma.

"I've given up all my other literary work. I'm no longer writing novels or short stories or plays. I sit working on nothing else but this. I can't tear myself away from it. I dreamed for so long about being able to study those lives. I collected these books. I didn't keep them together. I'd scatter them around so that a visitor couldn't see I'd collected *those* authors. So many writers passed through Kolyma! Hundreds, if not thousands. I collect everything I can find about them." Sometimes he has found confiscated poems and manuscripts in the archives as well.

Something more personal is also fueling Biryukov's quest. In the archives in Magadan, he recently found, complete with reports from informers, the case of his own father. His father was arrested here in Kolyma, where he had come to work as a factory manager. "I was born the day after my mother was allowed to hand in a parcel for my father for the first time. This was in October 1938; he had been arrested in May. Before that day, she hadn't heard any information about him. She was terribly upset when she finally did hear, and gave birth to me two months ahead of time." As an infant, Biryukov was taken back to "the mainland" by his mother.

Biryukov's father survived, and was released after a year and a half in prison. He rejoined his family for a short time—then, surprisingly, he returned to work in Kolyma. "I don't really know what he did in those years." Whatever it was, it occurs to me later, even as a "free" factory manager Biryukov's father would have been supervising convict labor, as capable as any NKVD officer of saving a life or working a prisoner to death. Biryukov does not mention it, but I wonder afterward if one source of his obsession may be an attempt to figure out how complicit his father was, before his arrest and after his release, in the slave labor economy of Kolyma. Was he more slave or slave driver? And was it a compulsion to answer this question that made Biryukov return to Magadan after finishing his education?

"I came to see the town where I was born," he says. But he has stayed, just as his father did. Whatever he is searching for, he has not found. "I'm not talking too much, am I? I can talk about all this endlessly. It's my madness, if you will."

In his "madness," this tall, intense man is a representative of a whole generation of people trying to find out whether their parents were victims or executioners—or both. Like a magnet, the greatest of the Soviet killing grounds has drawn Biryukov back. In fact, if any one spot could be called the center of that inferno, it is the spacious public square in the middle of Magadan where, on a park bench, he is telling his story. The buildings on several sides of this square once housed the Kolyma camps administration, other NKVD offices, courts, a prison, and an interrogation center. Guards could herd groups of prisoners through an underground tunnel linking several of the buildings. In this century, more human misery was directly administered from this square than from anywhere else on earth, except possibly the arrivals platform at Auschwitz, where Dr. Mengele decided who would live and who would die. For it was here that NKVD officials signed the lists approving where each Kolyma prisoner was to be sent. Would it be to a life-saving job in a dining hall or hospital? Or to a place like Camp Expeditsiony, near the Arctic Circle, where from one group of six hundred prisoners sent there in 1939, one hundred survived the first winter, and only one was alive six years later?

Biryukov is just as curious about my motives as I am about his. "And you," he asks, turning toward me on the bench. "Why are *you* so interested in all this?"

For the last time in Russia, I try to answer this question. One answer, I tell him, has to do with how a whole society, like an individual human being, must look deep into the past and face the very worst, in order to become healed. It is this that gives Russia's long, slow recovery a moral echo that goes beyond the experience of Stalinism.

Another answer has to do with the line between victims and executioners, far less distinct than I had first imagined. *You would have done the same.* It was this uneasy feeling that Eugenia Ginzburg took from her seventeen years of prison, labor camp, and exile: "After all, I was the anvil, not the hammer. But might I too have become a hammer?"

And then there is the equally indistinct line between that necessary Utopian wish to improve and remake the world and the wish for total power. The finger on the map: I want a railway— *there.* The impulses that can lead a person, or a whole society, to good or to evil, are not so distant from each other. To Biryukov I cite Milan Kundera, an anti-Communist writer if there ever was one, observing just this paradox in his native Czechoslovakia: ". . . the Communists took power not in bloodshed and violence, but to the cheers of about half the population. And please note: the half that cheered was the more dynamic, the more intelligent, the better half."

At a conference not long before I went to Russia, I heard a long-time member of the American Communist Party sadly try to explain why he had wanted to have faith in the Soviet Union all these years. "You know," he said, "I think it was because I so much wanted there to be somewhere, somewhere in the world, where they would do everything *right*." The danger comes when we leap to assume that this land of dreams must exist on earth, and not merely in our imagination; that Utopia is already there, in some other place, and cannot only be built slowly, brick by imperfect brick, at home. And then, making that fatal assumption, as did so many of the early pilgrims to the Soviet Union, we fail to see what is before our eyes. Millions of Soviets made the same mistake, refusing to see that the Revolution they had fought for was about to devour them. Yet Susanna Pechuro, a 17-year-old schoolgirl, saw everything. And so finally, I tell Biryukov, part of my own obsession has to do with the mystery of denial and of seeing.

On my last night in Magadan, I offer to take MM and his wife and daughter out to dinner. MM declines. "A lot of the people who work in the restaurants here are my patients, and they tell me what goes into the food!" He again invites me to his apartment. It is a mellow evening around the family table, with farewell toasts and laughter. Early tomorrow morning I will fly back, across eight time zones, to my family in Moscow; a few days later, we will return to the United States.

I say goodbye and walk back toward the hotel. But it is still light out, the long northern summer dusk, and I make a detour. MM's apartment is near the crest of the ridge that divides the main part of Magadan from its harbor, Nagayevo Bay. The evening air is clean and crisp, and the fog that has covered Magadan for most of the week has lifted. After a few minutes' walk I'm over the ridge and sitting on a wall with a stunning view down a long slope to the water, where ships are docked or ride at anchor. A freighter has just unloaded onto the wharf several dozen big yellow bulldozers, bound for the gold mines.

Beyond the docks stretches one of the great natural harbors of the Pacific: a long fjord, winding its way some miles from the shore below me out to the open sea, of which I can see a sunlit glimmer in the far distance. The fjord's high, steep sides shelter its ribbon of

water from the fierce storms of the Sea of Okhotsk. How thrilled
the early navigators must have been to find this place.

It is haunting to find such exhilarating beauty in the meeting
of land and water here, in this spot that was the gateway to so
much death. The contrast reminds me of another great slave port,
the Île de Gorée, off the coast of West Africa. Once the way station for most of the slaves who were brought to the New World,
branded and shackled, today it is a paradise of white beaches, coco-
nut palms, and whitewashed buildings drenched with red bougain-
villea.

It was through this harbor at Nagayevo Bay that almost all
prisoners came to Kolyma. Convicts who arrived in the spring,
when the tip of the bay was still frozen, had to march the last mile
or two over the ice. Vladimir Petrov, who arrived in Nagayevo Bay
as a prisoner on the ship *Dzhurma* in 1936, had a different experience that seemed filled with foreboding: the entire hillside of the
fjord was aflame with a forest fire. When Eugenia Ginzburg arrived here in 1939, she had been nearly unconscious for two days.
"One by one we invalids were carried ashore on stretchers and left
on the beach in tidy rows. The dead were also stacked nearby so
that they could be counted and the number of death certificates
would tally. Lying on the pebbly shore, we watched our comrades
being marched off toward the town. . . ."

Ginzburg was released in 1947 after ten years in prison and labor camp, was rearrested, then released again. She had to live and
work as an exile in Magadan until the mid-1950s. And for much of
the time she lived right here, in the part of town that still stands on
this hillside rolling down before me to the harbor: a shantytown of
small, ramshackle wooden houses, crammed up against each other,
mostly built by newly freed prisoners in the 1940s and 1950s. Wisps
of smoke drift upward from their chimneys. The district is known
locally as "Shanghai."

It was while living in Magadan that Ginzburg realized she had
survived, not only physically, but emotionally. It was here that she
could live at last with the man she fell in love with in the camps,

a doctor famous among both prisoners and guards for his gentle wisdom and his healing powers. It was here that she was at last allowed a summer visit from her sixteen-year-old son, from whom she had been separated when he was four. She had feared that she might have nothing in common with him—but he arrived carrying a knapsack filled with the books of poetry that were so precious to her. "... that very first night he started to recite from memory the very poems that had been my constant companions during my fight for survival in the camps." The scene of their meeting is all the more moving for Russians because they know that this teenager with his knapsack of books grew up to become the boldly dissident writer Vassily Aksyonov. And it was here that Ginzburg brought up another, adopted child, an orphan girl, whose coming gradually healed the ache from the death of her other son in World War II.

While living on this hillside above the harbor, Ginzburg had another, most unexpected, experience. It was soon after Stalin's death, and the authorities were just beginning to dismantle the *gulag*. Hope was in the Magadan air. The terrible witch craze at last seemed over. An exile, Ginzburg was still under the jurisdiction of an NKVD commandant, and had to report to him on the first and fifteenth of each month to get her papers stamped. She was not yet free to go back to "the mainland," but, to her joy, she was allowed to resume the work she had done before her arrest, as a teacher of Russian language and literature. She was assigned to teach an evening class at the Magadan adult school.

"Here are your pupils," said the director of the school, leading me into the classroom.

What was this? In front of me I saw officers with brilliant gold epaulettes and beautifully polished boots. One solid mass of officers. Forty of them. Among them I spotted faces I knew. These were our commandants! Both past and present, young ones and older ones. Later on it was explained to me that with change in the air officers were required to have reached a certain educational level; they had to hurry off to adult school to acquire the now essential graduation certificate.

Jarred by this reversal of roles, Ginzburg gingerly began teaching these NKVD officers about grammar and syntax. One evening, she slyly gave her students the example of a message from Tsar Nicholas II. Responding to an appeal from a prisoner sentenced to death, the Tsar telegraphed: "Execution, impossible reprieve." She showed how she could reverse the meaning by moving the comma one word to the right. "Now you see," she told the class, ". . . that a man's life may depend on a single misplaced or omitted comma."

The classes ran late at night. Vacant lots on this hillside above Nagayevo Bay were notorious hangouts for rapists and thieves. The officers in Ginzburg's class decided they would take turns escorting their teacher home each night, across the slope now spread out before me. "One day it was the turn of my own commandant, Gorokhov, to accompany me. The entire walk I kept on at him about the correct spelling of adjectival suffixes, and it wasn't until we were already going down the hill to Nagayevo that I suddenly remembered aloud:

" 'Oh yes, tomorrow is the fifteenth! So I'll be coming to see you in your office to get my documents stamped. . . .' "

The officer was embarrassed, groped for something to say, and finally spoke the same thought that was on Ginzburg's mind as well:

"I expect it'll soon be all over."

Like Ginzburg, so many of the people I've liked best in my journey across Russia have been, formally or informally, teachers: Igor Dolutsky, with his bold efforts to move beyond traditional schoolbooks; Susanna Pechuro, with the young people I saw riveted by her story; Vladimir Glebov, the philosophy professor who could laugh and joke about his own path through orphanage and prison; even the anguished schoolteacher Inna Sukhanova in Kolpashevo, struggling to come to terms with her father, teaching students who must include the grandchildren of people he ordered shot. If Russians can at last squeeze the slave out of themselves, as Chekhov wrote, it will be with the help of such men and women.

Forty years later, why is it so moving to remember that conversation between Eugenia Ginzburg and her NKVD commandant as they walk down this hillside toward the harbor? I think it is because of the way, in those first hesitant years of recovery from Stalinism, their roles are unexpectedly shifting. She is still his prisoner—out on probation, in effect—but she is also his teacher. He is still her jailer, but tonight he is also her protector from street criminals. No longer are they victim and executioner, witch and persecutor. They are each, partly, tentatively, transformed into something else.

Transformation was what the Russian Revolution was supposed to be about. The early Bolsheviks believed that human character itself could be changed if only the shape of the community we lived in were changed. And in a sense they were right, but in a far darker sense than they ever imagined. For the change, of course, can be for the worse. It is not only the empire of a Tsar that can turn someone into an exploiter of other human beings, but every other kind of unlimited power as well. And this is so whether that unlimited power stems from the existence of a secret police, from the absence of democracy and of legal safeguards, or from the prevailing belief that right and wrong is to be decreed by someone or something beyond ourselves—Tsar or Party, dictator or corporation, guru or sacred texts.

When the witch craze swept Europe, there were peculiar little pockets, a city here, a duchy there, that it did not touch. A wise government or a local tradition of tolerance simply never let the madness get started in the first place. We all carry in us the embryos both of an executioner and of a teacher or healer; it is the communities we build for ourselves that call forth a little less of the one and a little more of the other. That this once-feared harbor of death is now only a peaceful fjord at sunset, and that a ship is unloading tractors instead of slaves, is testimony that such change is possible, even if in Russia it is still uncertain and incomplete, filled with setbacks and wrong turnings. And if it can begin here, in Kolyma, perhaps it can happen anywhere.

This well-known photo served as a model for a Moscow statue inscribed, "Thank you, Comrade Stalin, for my happy childhood." Some two years after the picture was taken, the father of this six-year-old girl was shot. Her mother died soon after.

Bibliography and Acknowledgments

The shelf of books on the Stalin-era Soviet Union is an enormous one. For decades, many novels, stories, and memoirs were smuggled out and published in the West; in the last few years hundreds more, long locked in desk drawers, have been pouring off the presses in Russia itself. In history and biography, because of so many years of Soviet censorship, far more good work is available in English than in Russian. What follows is in no way a comprehensive bibliography on the subject, nor is it more than a fraction of the books and journals in both languages that I was able to consult, in the splendid library of the University of California at Berkeley. Instead, this list is limited to works in English, to those of more interest to a general reader than a Soviet specialist, and, in most cases, to books that are still in print.

Of the historians who have devoted their lives to the Stalin period, the work of two, above all, stands out: Robert Conquest in the United States and Roy Medvedev in Russia. Conquest's masterwork, revised in 1990, is *The Great Terror: A Reassessment*; Medvedev's is *Let History Judge: The Origins and Consequences of Stalinism* (revised and expanded edition, New York, 1989). These latest editions make use of new sources available since *glasnost*. Conquest is a conservative Englishman, Medvedev a Russian Marxist. Although they have differences of emphasis and interpretation, their pictures of Stalin's Russia are broadly similar. Medvedev's many other books include *All Stalin's Men: Six Who Carried Out the Bloody Policies* (New York, 1984). Of particular interest among Conquest's other works are *Kolyma: The Arctic Death Camps* (New York, 1978) and *The Harvest of Sorrow: Soviet Collectivization and the Terror-Famine* (New York, 1986), one of the remarkably few books about something that may have killed as many as 10 million people.

Of the hundreds of memoirs about Russia under Stalin, those of at least three people are works of lasting literature: Eugenia Ginzburg, Nadezhda Mandelstam, and Victor Serge. Ginzburg was an ardent Communist and the wife of a high provincial Party official when she was arrested in 1937. Her *Journey into the Whirlwind* (New York, 1967) describes her interrogation, solitary confinement, and the harrowing journey to Kolyma; *Within the Whirlwind* (New York, 1981) is about her survival there. Nadezhda Mandelstam, nearly a decade younger than her famous poet husband, was first a star-struck admirer, then virtually his secretary. But in the end, after his arrest and death, and after she fought a decades-long battle to preserve the texts of his poems, and to record their ordeals together and then hers alone, she proved herself as great a writer as he. Her *Hope Against Hope* (New York, 1976) and *Hope Abandoned* (New York, 1981) are an extraordinary love story, testimony to the power of steel-hard conscience, and the greatest portrait gallery we have of the many varieties of compromise and illusion in Stalin's Russia.

The accounts of these two women are well known; Victor Serge's *Memoirs of a Revolutionary* (New York, 1984) is a neglected classic. Although a minor Comintern official, Serge was at heart an anarchist and democrat who never fit comfortably into any political party. He quickly became part of the opposition to Stalin, was arrested and exiled, and then finally managed to get out of Russia just as the Great Purge began. In spirit and in the directness and power of his prose, he was akin to Orwell, who admired Serge's work and tried to help him find a British publisher. In a series of vivid sketches of characters and episodes, Serge's memoirs are a moving panorama of the terror that swept over the Soviet Union in the twenties and thirties, seen from the inside, with the eye of a fine novelist and with the passion of someone who had hoped the Russian Revolution would lead to something far different.

Hundreds of other people have written about their time in Soviet prisons and labor camps; one of the best books is *A World Apart*, by the Polish writer Gustav Herling (New York, 1951). The *gulag's* most deadly territory, Kolyma, has a whole literature of its own. Besides the works by Conquest and Ginzburg mentioned above, several memoirs are particularly graphic: *Magadan*, by Michael Solomon, a Romanian (Princeton, N.J., 1971); *Eleven Years in Soviet Prison Camps*, by Elinor Lipper, a Dutch woman (Chicago, 1951); and *Dear America! Why I Turned Against Communism*, by Thomas Sgovio, an American (Kenmore, N.Y., 1979). The two volumes of stories by the writer and poet Varlam Shalamov, *Kolyma Tales* (New York, 1980) and *Graphite* (New York, 1981), record this killing ground as dryly and powerfully as if Hemingway had been at Auschwitz.

Alexander Solzhenitsyn is probably more responsible than anybody for making the *gulag* known to the outside world, particularly through his autobiographical novels *One Day in the Life of Ivan Denisovich* (New York, 1963) and *The First Circle* (New York, 1968). *The Gulag Archipelago, 1918–1956: An Experiment in Literary Investigation* (three volumes; New York, 1974, 1975, 1978) is an encyclopedia of the whole system, compiled under extraordinarily difficult conditions, when the author was being constantly harassed by the KGB. It is marred, unfortunately, by a hectoring, sarcastic tone, which is unnecessary; the material is damning enough by itself.

Considering the *gulag* inevitably leads one to the German concentration camps. Terrence des Pres's *The Survivor: An Anatomy of Life in the Death Camps* (New York, 1976) is devoted entirely to this comparison. The many books of Primo Levi, one of the wisest and most eloquent survivors of the Nazi camps, touch on the topic too. But a comparison of these mass murders with those in this century elsewhere in the world—especially the little-remembered genocide by the Belgian conquerors of the Congo—remains, so far as I know, yet to be written.

Curiously, one of the best works about the early decades of the

Soviet Union is a long biography of its greatest loser, Leon Trotsky, by Isaac Deutscher: *The Prophet Armed: Trotsky, 1879–1921* (New York, 1954), *The Prophet Unarmed: Trotsky, 1921–1929* (New York, 1959), and *The Prophet Outcast: Trotsky, 1929–1940* (New York, 1963). The complex picture of Trotsky as a character of Shakespearean dimensions transcends the limitations of Deutscher's own orthodox Trotskyist outlook. Throughout his account of this brilliant, harsh, and ultimately tragic figure, Deutscher constantly wrestles with the question: Given Russia's heritage, could the Revolution have come out differently?

There are many biographies of Stalin, including a recent, short one by Robert Conquest, *Stalin: Breaker of Nations* (New York, 1991). Alex De Jonge's *Stalin and the Shaping of the Soviet Union* (New York, 1986) is particularly readable. Robert Tucker's will surely be the most definitive; two volumes have appeared so far: *Stalin as Revolutionary, 1879–1929: A Study in History and Personality* (New York, 1973) and *Stalin in Power: The Revolution from Above, 1928–1941* (New York, 1990). Much of the more theoretical recent academic writing about Stalinism is arid and unsatisfying; an exception is Moshe Lewin's various books, particularly *The Making of the Soviet System: Essays in the Social History of Interwar Russia* (New York, 1985).

The topic of Westerners who saw the Soviet Union through rosy glasses in the 1920s and 1930s is covered in Paul Hollander's *Political Pilgrims: Travels of Western Intellectuals to the Soviet Union, China and Cuba, 1928–1978* (New York, 1981) and in David Caute's *The Fellow Travelers: Intellectual Friends of Communism* (revised and updated edition, New Haven, Conn., 1988). Malcolm Muggeridge, who was a newspaper correspondent in Moscow in the 1930s, unforgettably describes such visitors in his autobiographical *Chronicles of Wasted Time* (New York, 1940). But for accounts of the rare early travelers who saw so clearly, like Bertrand Russell and Emma Goldman, one must turn to their own writings.

During the long years of official silence about Stalinism, from the early sixties to the late eighties, there was still a lively debate on the subject in Russia's *samizdat* press. Some of this material is especially well presented in the anthology *An End to Silence: Uncensored Opinion in the Soviet Union*, edited and with introductions by Stephen F. Cohen (New York, 1982). For a taste of the official line given to the public while the real discussion was going on underground, browse through the English-language edition of the massive *Great Soviet Encyclopedia*, which can be found in many libraries.

A number of new books deal with the reconsideration of Stalinism and Soviet history generally during the Gorbachev period. Among them are Alec Nove's *Glasnost in Action: Cultural Renaissance in Russia* (revised edition, Cambridge, Mass., 1990); *Stalin: The Glasnost Revelations*, by Walter Laqueur (New York, 1990); and *Soviet History in the Gorbachev Revolution*, by R. W. Davies (Bloomington, Ind., 1989). The entire August/September 1991 issue of the London-based magazine *Index on Censorship* is devoted to manuscripts and interrogation records, newly released from the KGB files, of writers arrested in the Great Purge.

David Remnick's *Lenin's Tomb: The Last Days of the Soviet Empire* (New York, 1993) is the best of the more general recent books on Russia, particularly sensitive to how the opening up of the past helped bring down the whole crumbling Communist edifice.

None of the recognized systems for transliterating the Cyrillic alphabet into the Roman one are completely satisfactory; it is a bit like fitting a creature with four legs into pants made for one with two. Furthermore, our normal ways of writing familiar Russian words are full of inconsistencies: we refer to the Bolsh*oi* Theater, but to Tolst*oy* the writer, even though the Russian ending of each word is the same. In transcribing Russian names, I have tried to follow whatever is the most common usage, even if it doesn't follow one of the official transliteration systems; hence, Gorky instead of Gor'ki, Trotsky instead of Trotskii, and so on. One exception: When a writer has appeared in English, I've spelled his or her name as published. Eugenia (Ginzburg), for example, is an Anglicization of the Russian Yevgenia. And her son Vassily Aksyonov's publisher spells the author's first name differently than Vasily

Grossman's does his. Similarly, in referring to place names, I have usually followed the spelling of English-language atlases, even when this seems eccentric, as with the Kolyma town of Galimyy. However, English-language maps that have appeared since the Soviet Union's breakup sometimes have changed the spelling of names in the former Soviet republics other than Russia. For example, Karaganda has now become Qaraghandy, from the Kazakh. But Karaganda it was during the time of the *gulag,* and Karaganda it still is to its largely Russian-speaking inhabitants, and so I've left it—and several other cities—the old way.

Many people helped make this book possible. First of all, my family, who so willingly moved to Moscow in the dead of winter. If my wife, Arlie, had interviewed the people in these pages, they would have revealed twice as much. As it was, she always pushed me to ask more, and to explore every idea to its limits. And, as always, she was my best manuscript critic. Our son Gabriel left a school near the beach in sunny San Francisco for one in Moscow that he had to walk to when it was still dark out, in snowstorms, when the temperature was well below zero. Our son David helped us settle into our apartment before he had to return to the U.S. to college.

Tanya Pogossova spent hundreds of hours putting onto paper almost all the interviews in this book. She took my tapes of conversations in Russian with more than sixty people, usually of anywhere from one to three hours each, and transcribed them directly into near-perfect English, then meticulously went through each tape and checked her translation. While we were in Moscow, she helped in many other ways as well. As she continued to transcribe a backlog of these interviews after my departure from Russia, she sent me, with each one, her own thoughtful comments about each transcript, pointing out things I otherwise would have missed, setting people in their social and generational context. As this work came to an end, Oleg Svetlov joined her in transcribing some of the final tapes.

Others in Moscow helped in various essential ways, among them

Andrei Kolesnikov, Misha Shevelov, Lev Gushchin, Olga Vronskaya, Dulce Murphy, and Gennady Alfrenko. The writer Vladimir Zapetsky generously shared with me the results of his own extensive research into the events at Kolpashevo. Vitaly Shentalinsky provided much useful advice and an inside look at his own remarkable project of getting the literary manuscripts of Great Purge victims released from the KGB archives. Kathleen Smith and Vladimir Klimenko not only steered me to some of the most interesting people I interviewed, but also later read the manuscript.

I'm grateful to all the people who agreed to let me interview them. Some opened many other doors as well; my debt to Wilhelm Fast in Tomsk and Miron Atlis in Magadan is, I hope, obvious. In the book, there was room for only a small proportion of the people I talked with, but my thanks go equally to the others; I could easily fill another entire book with their insights. I learned much, especially, from many hours with historians and researchers, most of them members of Memorial: Arseny Roginsky, Nikolai Formosov, Nikita Petrov, and many others in Moscow; Andrei and Irina Resnikov in Saint Petersburg; Vladimir Sirotinin in Krasnoyarsk; Nikolai Kashchaev in Tomsk; Asir Sandler in Magadan; and Sergei Krasilnikov, Vladimir Shishkin, Irina Pavlova, and Nikolai Pokrovsky, all of the Institute of History, Philology, and Philosophy in Novosibirsk. The filmmakers Leonid Gurevich, Arkady Kordon, Yevgeny Tsimbal, and Tofik Shakhverdiev all shared their thoughts with me and arranged for me to see films of theirs about the Stalin era or portions of works in progress. And a number of former *gulag* prisoners, besides those quoted in the book, shared their experiences with me: Boris Lesniak and Galina Lewinsohn in Moscow, Leonid Trus in Novosibirsk, Ivan Yegorov in Krasnoyarsk, and Vasily Kovalev in Magadan. And, for many memorable conversations over the years about all the issues in this book and much more, first in Moscow, and later in Germany and the United States, my thanks and admiration to Lev Kopelev and the late Raisa Orlova.

In Novosibirsk, Wendy Bewig and her colleagues at the Phys-Mat School were our hosts for a week, and Nikolai Shelyaev arranged a number of interviews. In Tomsk, Yura Kudinov put me up in his apartment for a week when the city's hotels were full; he and his friends helped make the middle of Siberia feel like home.

Back in the United States, special thanks to Nan Graham, my editor, and Georges Borchardt, my agent, for their comments on the manuscript; to Denise Shannon, for more agenting help, and to Beena Kamlani, for shepherding the book into production; and to Jean Kilbourne, who suggested the elephant-in-the-living-room analogy. Ellen Binder took many of the pictures that appear in these pages, and David King, Carol Leadenham and Sondra Bierre helped find still more. Barbara Jackson worked long and hard on the map. Hunter Pearson kept an eye on things at home while we were away, and without Joel Schatz we wouldn't have had a roof over our heads in Moscow.

A *bolshoe spasibo* to all those who labored so hard to teach me Russian over the years, especially Veronica Dolenko, Valya Perelotova, Svetlana Peterburgskaya, Sima Radivilova, and Zhenya Sokolov and the staff of the Norwich University Russian School. And to my late father, in whose company I first started studying this impossible language some thirty years ago. He always said that someday I'd be glad I made the effort, and he was right.

Finally, I was blessed with a number of friends, beside those already mentioned, who read and commented on the manuscript. Nothing in the process of writing did I look forward to more than their criticisms and suggestions, whether these came over a cup of coffee in San Francisco or by electronic mail from Moscow. And so, my warmest thanks to Harriet Barlow, Peter de Lissovoy, Tom Engelhardt, Doug Foster, Todd Gitlin, Richard Greeman, Barbara Harmel, Hermann Hatzfeldt, David Hochschild, Ilse Jawetz, Judith Klein, Elinor Langer, Cindy Scharf, and Allen Wheelis, and to Mary Felstiner, who brought to the reading of two successive drafts the benefit of her own extraordinary work on the Holocaust. A red-penciled manuscript, filled with crossed-out passages and question marks in the margins, is the greatest gift a writer can receive, for it makes his work not so solitary after all.

Index

Page numbers in italic type indicate illustrations.

ABOUT THE AUTHOR

Adam Hochschild was born in New York City in 1942. His first book, *Half the Way Home: A Memoir of Father and Son,* was published in 1986. "By turns nostalgic and regretful, lyrical and melancholy," wrote Michiko Kakutani of the *New York Times,* "*Half the Way Home* creates . . . an extraordinarily moving portrait of the complexities and confusions of familial love . . . conjuring them up with Proustian detail and affection." It was followed by *The Mirror at Midnight: A South African Journey* and *The Unquiet Ghost: Russians Remember Stalin. The Unquiet Ghost* won prizes from the Overseas Press Club of America and the Society of American Travel Writers.

Hochschild's *Finding the Trapdoor: Essays, Portraits, Travels* won the PEN/Spielvogel-Diamonstein Award for the Art of the Essay. *King Leopold's Ghost: A Story of Greed, Terror, and Heroism in Colonial Africa* was a finalist for the National Book Critics Circle Award. It also won a J. Anthony Lukas Prize, the Duff Cooper Prize in Great Britain, and the Lionel Gelber Prize in Canada. Hochschild's books have been translated into ten languages.

Besides his books, Hochschild has written for *The New Yorker, Harper's Magazine,* the *New York Review of Books,* the *New York Times Magazine, Mother Jones,* the *Times Literary Supplement,* the *London Review of Books,* and many other newspapers and magazines. He is a former commentator on National Public Radio's *All Things Considered.*

Hochschild teaches writing at the Graduate School of Journalism of the University of California at Berkeley and has been a guest teacher at other campuses in the United States and abroad. He spent five months as a Fulbright lecturer in India. He lives in San Francisco with his wife, Arlie, a sociologist and author. They have two sons.

KING LEOPOLD'S GHOST

A STORY OF GREED, TERROR, AND HEROISM IN COLONIAL AFRICA

A New York Times Notable Book

A National Book Critics Circle Award finalist

"An enthralling story, full of fascinating characters, intense drama, high adventure, deceitful manipulations, courageous truth-telling, and splendid moral fervor."

— *Christian Science Monitor*

Beginning in the 1880s, King Leopold II of Belgium carried out a brutal plundering of the territory surrounding the Congo River. Ultimately slashing the area's population by an estimated ten million people, he still managed shrewdly to cultivate his reputation as a great humanitarian. *King Leopold's Ghost* is the haunting account of a megalomaniac of monstrous proportions. It is also the deeply moving portrait of those who fought Leopold: a brave handful of African rebels, missionaries, and young idealists who participated in the twentieth century's first great international human rights movement.

ISBN 0-618-00190-5